netics fo   icologists

The Remedica Genetics for... Series
Genetics for Cardiologists
Genetics for Dermatologists
Genetics for Hematologists
Genetics for Oncologists
Genetics for Ophthalmologists
Genetics for Orthopedic Surgeons
Genetics for Pulmonologists
Genetics for Rheumatologists

Published by the REMEDICA Group

REMEDICA Publishing Ltd, 32–38 Osnaburgh Street, London, NW1 3ND, UK
REMEDICA Inc, Tri-State International Center, Building 25, Suite 150, Lincolnshire, IL 60069, USA

E-mail: books@remedica.com
www.remedica.com

Publisher: Andrew Ward
In-house editors: Tamsin White & Eilidh Jamieson
Design: REGRAPHICA, London, UK

ISBN 1 901346 19 6
British Library Cataloging-in Publication Data
A catalogue record for this book is available from the British Library

Printed at Ajanta Offset & Packagings Limited, India.

Genetics for Oncologists
The molecular genetic basis of oncologic disorders

Fiona Lalloo
Consultant in Clinical Genetics,
St Mary's Hospital,
Manchester
UK

Series editor:
Eli Hatchwell
Investigator
Cold Spring Harbor Laboratory

REMEDICA
publishing

LONDON • CHICAGO

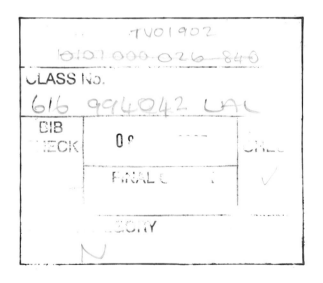

Introduction to Genetics for... series

Medicine is changing. The revolution in molecular genetics has fundamentally altered our notions of disease etiology and classification, and promises novel therapeutic interventions. Standard diagnostic approaches to disease focused entirely on clinical features and relatively crude clinical diagnostic tests. Little account was traditionally taken of possible familial influences in disease.

The rapidity of the genetics revolution has left many physicians behind, particularly those whose medical education largely preceded its birth. Even for those who might have been aware of molecular genetics and its possible impact, the field was often viewed as highly specialist and not necessarily relevant to everyday clinical practice. Furthermore, while genetic disorders were viewed as representing a small minority of the total clinical load, it is now becoming clear that the opposite is true: few clinical conditions are totally without some genetic influence.

The physician will soon need to be as familiar with genetic testing as he/she is with routine hematology and biochemistry analysis. While rapid and routine testing in molecular genetics is still an evolving field, in many situations such tests are already routine and represent essential adjuncts to clinical diagnosis (a good example is cystic fibrosis).

This series of monographs is intended to bring specialists up to date in molecular genetics, both generally and also in very specific ways that are relevant to the given specialty. The aims are generally two-fold:

(i)   to set the relevant specialty in the context of the new genetics in general and more specifically

(ii)   to allow the specialist, with little experience of genetics or its nomenclature, an entry into the world of genetic testing as it pertains to his/her specialty

These monographs are not intended as comprehensive accounts of each specialty — such reference texts are already available. Emphasis has been placed on those disorders with a strong genetic etiology and, in particular, those for which diagnostic testing is available.

The glossary is designed as a general introduction to molecular genetics and its language.

The revolution in genetics has been paralleled in recent years by the information revolution. The two complement each other, and the World Wide Web is a rich source of information about genetics. The following sites are highly recommended as sources of information:

1.   PubMed. Free on-line database of medical literature.
     http://www.ncbi.nlm.nih.gov/PubMed/

2.   NCBI. Main entry to genome databases and other information
     about the human genome project.
     http://www.ncbi.nlm.nih.gov/

3.   OMIM. On line inheritance in Man. The On-line version of
     McKusick's catalogue of Mendelian Disorders. Excellent links
     to PubMed and other databases.
     http://www.ncbi.nlm.nih.gov/omim/

4.   Mutation database, Cardiff.
     http://www.uwcm.ac.uk/uwcm/mg/hgmd0.html

Eli Hatchwell
Cold Spring Harbor Laboratory

The genetic basis of cancer

Introduction

Cancer genetics is now one of the fastest expanding specialties and accounts for about 40% of referrals to regional genetics departments within the UK. In the last few years, the genetic basis of some of the common malignancies – such as breast, ovary and colorectal cancer – have been partially elucidated, allowing the targeted screening and management of patients with a cancer predisposition.

One of the premises for the study of relatively rare inherited cancers is the assumption that understanding the molecular basis of these malignancies will result in better understanding and treatment strategies of the more common malignancies. While this has not yet resulted in new therapies, the management at diagnosis of many patients with an inherited predisposition has been radically altered. Thus, a working knowledge of cancer genetics will allow the clinician not only to appropriately manage patients, but also to comprehend the major advances in the field of oncology.

At the molecular level, cancer is caused by mutations in DNA, which result in aberrant cell proliferation. Most of these mutations are acquired in an age-dependent manner and occur in somatic (uninherited) cells. However, some people inherit mutations in the germline (the gametes and their precursors).

Inherited or genetic forms of cancer may be due to mutations in genes that directly control cell growth or apoptosis, which have been termed 'gatekeeper genes' by Kinzler and Vogelstein (Nature 1997;386:761–3). These include genes that control cell growth – proto-oncogenes – and genes that have a negative effect on cell growth – tumor suppressor genes. Conversely, the inherited mutation may be in a gene that indirectly influences the cell cycle or indirectly increases the mutation rate in cells controlling cell growth by altering factors within pathways related to cell cycle control. These are known as 'caretaker genes'. Many caretaker genes appear to act in pathways that involve gatekeeper genes.

Oncogenes

Oncogenes are derived from normal cellular genes called proto-oncogenes. Proto-oncogenes were first elucidated in RNA tumor viruses, and are now known to encode proteins that are crucial for normal cellular growth regulation including growth factors, components of the intercellular signaling pathways, DNA binding proteins, cell surface receptors and components of the cell cycle progression pathways.

The transformation of a proto-oncogene to an oncogene is the result of a gain in function and can occur in a number of ways: overexpression of the gene, or amplification (such as a duplication) to produce increased oncoprotein; activation or formation of fusion genes by translocation; or alteration of the gene product to produce transforming proteins.

As well as initiating an oncogenic process, oncogenes are thought to be important in the maintenance of solid tumors. Inherited mutations of oncogenes are rare—it is thought that most of these would be lethal. However, mutations in the *RET* oncogene and the *MET* oncogene have been described and cause inherited susceptibility to malignancy. (See sections on Multiple endocrine neoplasia type 2 and Familial papillary renal cell carcinoma.)

Tumor suppressor genes

Under normal circumstances, tumor suppressor genes regulate cellular differentiation and suppression of proliferation. Mutations in these genes result in unchecked cellular proliferation, resulting in tumors with abnormal cell cycles and tumor proliferation. Pathogenic mutations in tumor suppressor genes act by loss of function as opposed to the gain of function mutations in proto-oncogenes.

Knudson put forward a model of how mutations in tumor suppressor genes may result in malignancy following statistical studies of retinoblastoma – Knudson's two-hit hypothesis of

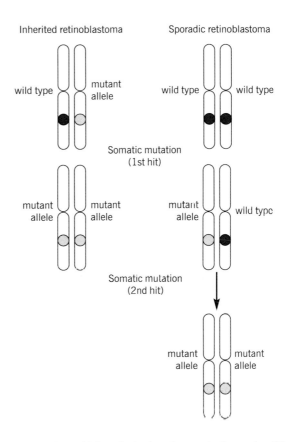

Figure 1. Knudson's two-hit hypothesis of carcinogenesis. Two copies of the mutant allele result in loss of function of the gene and altered cell proliferation. This occurs earlier (after only one 'hit') in the inherited form of retinoblastoma.

carcinogenesis (Proc Natl Acad Sci 1971;68:820–3). Knudson observed that patients with inherited retinoblastoma were more likely to have bilateral disease occurring at a younger age than patients with sporadic disease. He postulated that malignancy followed mutations in two alleles at the same locus and was a two-step process. In familial retinoblastoma, one mutation is inherited in the germline, therefore, only one further mutation is necessary for tumor development. However, in sporadic tumors, two separate events are required to develop a malignancy. Because the

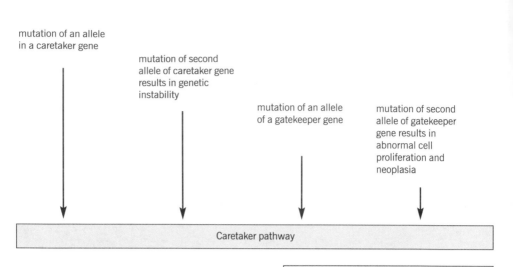

mutation of an allele
in a caretaker gene

mutation of second
allele of caretaker gene
results in genetic
instability

mutation of an allele
of a gatekeeper gene

mutation of second
allele of gatekeeper
gene results in
abnormal cell
proliferation and
neoplasia

Caretaker pathway

Gatekeeper pathway

NORMAL CELL   CANCER

Figure 2. The effect of mutations in caretaker and gatekeeper genes.
Adapted from Kinzler KW, Vogelstein B: Nature 1997;386:761–3.

probability of acquiring a single mutation is much greater than acquiring two mutations, malignancies would develop at earlier ages, and would be more likely to be multiple in those people with a germline mutation than in patients with sporadic disease (see Figure 1).

Caretaker genes

Caretaker genes act in a pathway involved in the regulation of the cell cycle and cell differentiation, but do not directly affect the cell cycle. Caretaker gene mutations do not directly result in tumorigenesis, but result in loss of function, which leads to genetic instability and an increased mutation rate in other genes including gatekeeper genes.

According to Knudson's hypothesis, gatekeeper genes need two mutations to become malignant; however, this hypothesis is modified with caretaker genes. It is suggested that an inherited germline mutation in a caretaker gene would require a further three acquired mutations to cause a malignancy: a mutation in the remaining allele of the caretaker gene and mutations in both alleles of the gatekeeper gene (see Figure 2). The statistical probability of acquiring three further mutations is lower than acquiring one further mutation, therefore, the incidence of malignancy resulting from germline mutations in caretaker genes is likely to be lower than that resulting from germline mutations in gatekeeper genes.

For example, BRCA1 (one of the genes implicated in hereditary breast cancer) is known to be associated with Rad51, a protein involved in double-stranded DNA repair. BRCA1 is also thought to be involved in the control of mitotic spindles. Disruption of this gene may then initiate genetic instability due to the accumulation of DNA damage and chromosomal abnormalities, which would result in mutations in other genes. BRCA1 has been implicated in the p53 pathway (a ubiquitous pathway controlling apoptosis-programmed cell death) and it is possible that mutations in BRCA1 may eventually lead to mutations in Tp53, a gatekeeper gene. The role of genetic instability in tumorigenesis remains under investigation.

This model of mutation would account for the reduced penetrance seen in some inherited cancer syndromes. For example, mutations in BRCA1 have a penetrance of about 85%, in comparison, mutations in TP53 have a penetrance approaching 100%.

Other examples of caretaker genes include the DNA mismatch repair genes, which cause hereditary non-polyposis colorectal cancer, and the nucleotide excision repair genes causing xeroderma pigmentosum.

Counselling for inherited predisposition to cancer

The aim of genetic counselling is to inform patients and their families about their genetic predisposition, in order to facilitate understanding and potential decision-making about management options. Construction of a three-generation pedigree allows accurate diagnosis/assessment of risks (see **Glossary** for pedigree drawing and explanation).

There are a number of general themes that apply to the counselling of families with a genetic predisposition to cancer.

- **Allocating sufficient time** to take an accurate family history and allow discussion of issues is fundamental to the counselling process.

- It is important to **assess the patient's reason for attending the clinic**, whether to be informed about their risk, to talk about options with regard to prevention and screening or, as is often the case, simply because they have been told to do so by another health care professional. This latter group of patients may be the most difficult to counsel as they are often unprepared for any information regarding an increased risk.

- Most **individuals wish to define their risks** of developing a particular form of cancer. However, the concept of risks, both absolute and relative, is often difficult for patients to grasp. Many find numerical risk figures difficult to understand and prefer simply to know whether or not they are at increased risk of developing a condition.

- In dealing with individuals with a family history of cancer, an individual's **perception of risk is colored by personal experience of the disease**. For example, a young woman whose mother died of breast cancer may find being told that she has only a moderate risk of the disease much more upsetting than a woman who is at high risk but from a family where most of the women have survived with breast cancer.

- **Individuals often seek advice shortly after the death of a loved one/relative**. This is a particularly difficult time when individuals are not only anxious about their own risk, but are still working through the grieving process. During this stage of bereavement, it is advisable that individuals are encouraged not to make any irreversible decisions, for example regarding predictive testing or preventative surgery.

- When **individuals approach the age at which a relative was diagnosed with cancer** they become aware of their own mortality and may initiate appointments with the intention of starting screening.

- As with most adult-onset conditions, **genetic testing is usually avoided in children** unless it would alter the management of those at risk. Whilst many parents would like to know whether or not their children have inherited a mutation that they themselves carry, most understand that predictive testing is a decision best left to the child once they have become an adult and are capable of making the decision themselves.

Counselling issues for many of the conditions described in this book are broadly applicable. As a result they are not discussed extensively in each section. However, it should be remembered that individuals will be anxious and need to be handled sensitively and, since the issues are very specific, differently to those with non-genetic forms of cancer. Patients need to be given as much information as possible and should be allowed to make their own decisions about whether or not to begin screening, proceed to genetic testing (if it is available) and/or actively pursue preventative measures such as surgery if applicable.

Genetic testing

An increasing number of genes are being cloned that are responsible for inherited cancer syndromes (e.g. von Hippel Lindau) or for autosomal dominant inheritance of common malignancies such as breast cancer. It is assumed that once a gene has been cloned, genetic testing will become available immediately; however, there are a number of factors to consider:

- Screening for mutations in a gene is currently a labor-intensive and time-consuming process. In most cases it requires analysis of the whole gene and may take a number of months.

- Mutation screening is usually only undertaken on blood samples

from an affected member of the family (DNA extraction from pathological blocks is still difficult, and poor quality DNA excludes screening of large genes). Mutation screening of an unaffected family member is only
of use once a family mutation has been found. Therefore, mutation screening is not available in a number of families.

- A negative result in an unaffected person may not always be conclusive. It may mean that a mutation is not present, or for technical reasons a mutation has not been detected, or, in the case of the more common malignancies, the mutation may be present in a currently unidentified gene.

Once a mutation has been identified within a family, predictive testing then becomes available for the unaffected members of that family. Predictive testing is only offered in this situation.

Predictive testing

In genetic centers in the UK, protocols have been devised for predictive testing in families with adult-onset Mendelian disorders. These are also widely used internationally. However, there are no nationally agreed guidelines within the USA. These UK guidelines are largely based on protocols for testing in families with Huntington disease. Such management procedures include discussion of the risks associated with mutations, the implications to the wider family of a mutation result, and exploration of how patients may cope with a result confirming that they are at high risk of developing a malignancy.

Patients are also advised to seek financial advice: the situation with regard to life insurance companies gaining access to results is currently under debate. It has been suggested that if an unaffected patient has had predictive testing and wishes to take out a new policy, insurance companies will have the right to access the result in order for them to calculate risk more accurately. However, a policy taken prior to a predictive test will remain valid. It has been agreed that insurance companies in the UK will be unable to demand predictive testing for a relative from a family with a known mutation. Legislation within the USA prevents health insurance companies from discrimination on the basis of genetic information.

Discussion of a potential 'good news' result includes reminding the patient that they would still have a population risk of developing a malignancy and that they would not be included in any screening programs for high risk patients.

In general, predictive testing protocols require three sessions of counselling. The first is an information session, discussing family history and potential risks of malignancy and the applicability of screening programs. The second session discusses predictive testing if available. A blood sample is drawn at the third session, which is at least a month later than the second. This delay ensures that the patient has sufficient time to consider the testing procedure and to debate the decision with appropriate members of family or friends. Both the second and third sessions are initiated by the patient and not by the genetics department. It is emphasized that testing can be done at any time in the future and can be delayed until a later date All patients are asked to attend clinic for a results session – only in exceptional circumstances would results be given by telephone or in a letter. At the beginning of the protocol for predictive testing, all patients are made aware of the need for three sessions and are given an approximate time span from the initial appointment to receiving a result.

Practicalities of cancer genetics

In the UK, the majority of molecular testing for inherited cancer syndromes is undertaken in the National Health Service (NHS) diagnostic laboratories. Most of these laboratories will only undertake a test if the patient is seen and counselled by a regional genetic center. This ensures that the test is appropriate, as well as ensuring that patients are fully aware of the implications of testing and have the opportunity to discuss these issues with professionals who are aware of potential pitfalls and are used to non-directive counselling. Regional genetic departments are happy to discuss whether or not potential referrals of high-risk families are appropriate. UK referral guidelines to regional genetic centers are available for the referral of patients with a family history of cancer.

Within the USA, clinical genetics centers do not operate in a cohesive network as in the UK, but *ad hoc* services have developed in most states. Therefore, genetic counselling is widely accessible. Individual laboratories offer genetic testing for a variety of disorders. These can be accessed through www.genetests.org

How to use this book

I have written this book in the hope that it will offer guidance in the diagnosis and management of inherited cancers. As with any book about molecular genetics, by the time of publication, some information will be out of date. However, those conditions where molecular genetic testing is available on a service basis will not change. EDDNAL (European directory of DNA laboratories www.eddnal.com) is a useful tool to determine which laboratories are offering testing of which syndromes.

The book is set out in three sections: inherited cancer syndromes, which are listed in alphabetical order; common cancers (with discussion of any specific genes known to be associated with these diseases), which are arranged according to the site of the malignancy; and chromosomal abnormalities in cancer.

Some gene mutations cause cancer syndromes with tumors in multiple sites and these clinical syndromes were described before the molecular genetics were known. In these cases I have described the syndromes. In other cases, mutations in certain genes are known to result in specific tumors and in these instances I have described the gene under the section of the site of malignancy.

Contents

1. Familial cancer syndromes

Ataxia telangiectasia

(also known as: AT)

MIM	208900
Clinical features	Characterized by progressive cerebellar ataxia (100%), oculocutaneous telangiectasia and immune defects. The ataxia develops within the first decade, and progresses to choreoathetosis (90%), dysarthria and oculomotor ataxia. Serum IgA levels are decreased or absent (60%), as are serum IgG2 levels (80%). About 10–20% of AT patients develop malignancies, especially leukemias and lymphomas. An increased incidence of gastric, ovarian, breast, liver and pancreatic malignancies have been reported. Chromosomal breakage is typical, and AT cells are abnormally sensitive to ionizing radiation.
Age of onset	Progressive ataxia from early childhood. Telangiectasia develops between 3–5 years of age.
Epidemiology	1 in 300,000
Inheritance	Autosomal recessive
Chromosomal location	11q22.3
Gene	*ATM* (ataxia telangiectasia mutated): 66 exons spanning 150 kb genomic DNA, encoding a 370 kDa protein.
Effect of mutation	The *ATM* gene contains a domain encoding a phosphatidylinositol 3-kinase (PI3-K) like protein. PI3-K proteins are involved in eukaryotic cell cycle control, DNA repair and DNA recombination. The ATM protein is known to have a role in the post-translational activation of p53. In the normal cell, ATM stabilizes p53 by phosphorylation, therefore a mutation of the *ATM* gene results in

unstable p53. ATM is also thought to be involved in the repair of double strand breaks, and is therefore important in the maintenance of the genome. Further work is required to elucidate all functions of ATM to account for the variable phenotype.

Diagnosis and counselling

Diagnosis can be confirmed by elevated levels of α-fetoprotein and carcinoembryonic antigen. Patients with AT are abnormally sensitive to ionizing radiation and therefore malignancies should not be treated with conventional doses of radiotherapy.

Mutation testing is available and prenatal diagnosis can be offered to families at risk of having a child with AT.

AT heterozygotes and breast cancer

There is controversy regarding the issue of ATM heterozygotes and breast cancer. It has been suggested that female AT heterozygotes have an increased risk of breast cancer and that it accounts for up to 18% of cases in US Caucasians. One study suggested that the risk of breast cancer is 6.1-fold higher in heterozygotes than in wild-type homozygotes. However, not all subsequent studies of women with early-onset breast cancer have supported these results. The most recent study suggests that heterozygotes are at increased risk of early-onset, bilateral disease, but with long-term survival. Larger studies are required to clarify the cancer risks associated with AT heterozygote status.

Basal cell nevus syndrome

(also known as: Gorlin syndrome; nevoid basal cell carcinoma syndrome [NBCCS])

MIM 109400

Clinical features Multiple basal cell carcinomas of the skin (BCCs) develop from nevi on sun-exposed skin. Metastases are rare. Multiple odontogenic keratocysts of the jaw are common; BCCs and jaw cysts occur in over 90% of patients by 40 years of age. The syndrome also manifests itself in calcification of the falx cerebri, bridging of the sella turcica, macrocephaly with biparietal frontal bossing, lateral displacement of the inner canthus, bifid and/or missing/fused ribs and abnormal segmentation of cervical vertebrae. Only 5% of patients develop medulloblastoma and this occurs at an average age of 2 years. There is an increased adverse response to radiation with locally destructive consequences. Therefore, radiotherapy should not be used to treat BCCs and should be reserved for second line treatment of medulloblastomas. Other complications include ovarian calcification or fibroma (24%), cardiac fibroma (3%), cleft palate (5%) and ophthalmological abnormalities (26%).

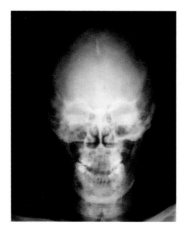

X-ray demonstrating calcification of the falx cerebri and odontogenic keratocysts.

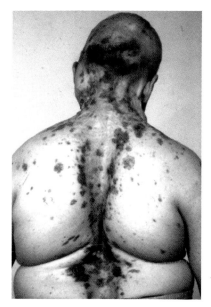

BCCs in the radiation field of treatment for a previous medulloblastoma.

Screening

NBCCS patients and those at risk of developing the syndrome are offered annual dermatological examination with open access if there is a suspicious lesion. Orthopantomograms are done on a 2-yearly basis.

Age of onset

BCCs usually develop around the age of puberty. The medulloblastomas present on average at 2 years of age, compared to 7.6 years in the general population.

Epidemiology

1 in 56,000

Inheritance

Autosomal dominant. 75% of patients develop BCC by 20 years of age, 90% by 40 years and 10% are non-penetrant.

Chromosomal location

9q22.3

Gene

PTCH: 23 exons spanning 34 kb, homologous to the Drosophila 'patched' gene. There are three different forms of the PTC protein, which may be due to the different splice forms of PTC mRNA. Somatic mutations in *PTCH* have been demonstrated in sporadic BCCs.

Effect of mutation

Hedgehog, a molecule influencing cell differentiation during development, acts via a receptor complex involving patched (PTCH) and smoothened (SMOH) transmembrane proteins. PTCH is the major receptor for all three hedgehog genes. It has been suggested that the hedgehog system is involved in differential signaling to basal cells throughout life and that the absence of PTCH may disrupt this signaling pathway. The wide variation in phenotype in BCNS may be due to other modifying genes.

Genetic testing

Mutation testing is now available but is time consuming. If a mutation is present within a family, predictive testing may be offered in order to direct screening, although diagnosis is often obvious clinically at a young age.

Beckwith–Wiedemann syndrome

(also known as: BWS)

MIM 130650

Clinical features Characterized by exophthalmos (80%), macroglossia (97%) and overgrowth (88%) in the neonate. Further features include neonatal hypoglycemia, visceromegaly, hemihypertrophy and linear indentations of the earlobes. There is an increased risk of malignancy in patients with BWS, generally occurring before 5 years of age. Potential malignancies include hepatoblastoma, adrenal cell carcinoma, rhabdomyosarcoma and Wilms tumor. The risk of neoplasia appears to be related to hemihypertrophy. About 12% of patients with BWS develop hemihypertrophy, and 40% of BWS patients with tumors have this feature.

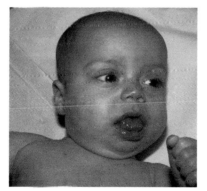

Macroglossia in a child with Beckwith–Wiedemann syndrome.

Age of onset From birth

Epidemiology 1 in 13,700

Inheritance Complex. Autosomal dominant forms with variable expressivity. Contiguous gene duplication at 11p15. Genomic imprinting disruption resulting from paternal uniparental disomy or defective or absent maternal BWS gene.

Chromosomal location	11p15.5
Gene	*CDKN1C* (Cyclin-dependent kinase inhibitor 1C, also known as p57kip2) encodes a 316 amino acid protein. Mutations are found in 5–20% of BWS cases.
	IGF2: overexpression of the paternally expressed gene. (See *imprinting* in the glossary.)
	H19
	KCNQ1OT1
Function of protein	All four genes lie within the same imprinted domain. *CDKN1C* and *IGF2* probably have a functional interaction. The *CDKN1C* gene is a negative regulator of cell proliferation. It is expressed from the maternal allele, and therefore, if the mutated gene is inherited from the father, there is no phenotypic abnormality. If inherited from the mother, the mutated gene is expressed and results in a BWS phenotype. Mutations have been demonstrated in *CDKN1C*, but abnormalities of imprinting of the other genes may account for the development of BWS.
	IGF2 is paternally expressed and shares an enhancing region with *H19*, which is maternally expressed. *KCNQ1OT1* is expressed normally from the paternal allele, but in BWS is abnormally expressed from both paternal and maternal alleles. Tumor formation in BWS appears to be more likely with abnormal methylation of the *H19* gene. Abnormal methylation usually results in aberrant expression of an imprinted gene.
Diagnosis and counselling	Neonates should be screened for hypoglycemia at the time of diagnosis. Abdominal examination and ultrasound should be performed in the neonatal period, then at 3-monthly intervals for the first 3 years, then every 6 months until the age of 8 years. Annual ultrasound should be considered until 12 years of age.

All patients with BWS should have a karyotype to check for translocations involving 11p15, and should have uniparental disomy studies. Mutation testing for *CDKN1C* is not available. Research studies are still ongoing to clarify the genetics of BWS.

Cowden disease

(also known as: multiple hamartoma syndrome)

MIM	158350

Clinical features Multiple hamartomas (benign disorganized growths) occur in numerous sites. Dermal lesions—trichilemmomas—are characteristic and occur in 99% of patients. Oral mucocutaneous lesions (82%), hamartomatous gastrointestinal (GI) polyps (35%), thyroid adenomas and multinodular goiters (40–50%), breast fibroadenomas (70%) and lipomas (42%) may also be present. Macrocephaly occurs in a proportion of patients and learning difficulties have been described, especially in association with Lhermitte–Duclos disease (dysplastic gangliocytoma of the cerebellum). There is an increased incidence of early-onset breast cancer (up to 50%) with a high bilateral risk and average age at onset of 38 years. An increased incidence of non-medullary thyroid cancer (7%) and endometrial cancer has been described along with reports of renal cell carcinoma.

Age of onset From birth onwards. The mucocutaneous changes tend to occur from 20 years, with breast cancer occurring in the third and fourth decade.

Epidemiology 1 in 200,000

Inheritance Autosomal dominant

Chromosomal location 10q23.3

Gene *PTEN* (phosphate and tensin homolog): 9 exons encoding a 403 amino acid dual specificity phosphatase. Mutations in *PTEN* have been described in Bannayan–Zonana (B–Z) syndrome, an autosomal dominant condition characterized by macrocephaly, intestinal polyps, multiple lipomas, pigmented macules of the glans penis and developmental delay. There is no documented increase in malignancy. B–Z syndrome is probably allelic with Cowden disease. No genotype/phenotype correlation has been found.

Somatic mutations in *PTEN* have been found in sporadic endometrial and thyroid cancers, malignant melanomas, glioblastomas and advanced prostate cancers.

Effect of mutation

PTEN is believed to negatively control the regulation of cell growth through the PI3-kinase/Akt signaling pathway. It is also involved in the control of cell migration, formation of focal adhesions and extracellular matrix interactions. It has been suggested that mutations in, or the absence of, PTEN may

- induce tumorigenesis by increasing cell proliferation

- reduce tumor cell apoptosis

- increase tumor invasiveness by decreasing anchorage dependence

Diagnosis and counselling

Breast examination and annual mammograms should be offered from 30 years of age—prophylactic mastectomy may be considered as an option. Screening for thyroid abnormalities is recommended. A recent report suggests annual endometrial screening from 30 years and annual urinalysis for blood. Mutation testing is not widely available at present.

Familial adenomatous polyposis coli

(also known as: FAP)

MIM   175100

Clinical features   FAP is characterized by very large numbers of adenomatous polyps
in the colon or rectum that progress to malignancy if untreated,
usually in the 4th decade (average age 35 years). Adenomatous
polyps also occur in the upper GI tract, this is a major cause of
morbidity for those patients in whom colorectal cancer is prevented.
Duodenal adenocarcinoma occurs in 12% of cases and gastric
adenomas predispose to gastric malignancy. Extracolonic features
include epidermoid cysts (66% of cases), osteomas of the mandible
(90%), dental anomalies and congenital hypertrophy of the retinal
epithelium (CHRPE). Desmoid tumors occur in 5–10% of cases and
are locally invasive, generally in the abdomen. Papillary carcinoma
of the thyroid, hepatoblastoma and medulloblastoma have all been
described with FAP.

An attenuated form of FAP (<100 colonic adenomas) has been
described with a later onset of colorectal cancer (>40 years of age).

Colorectal cancer
in a specimen
with multiple
polyps.

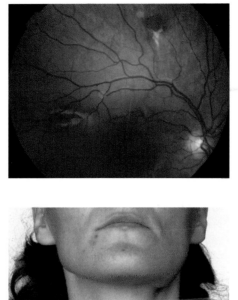

Multiple areas of CHRPE in FAP.

Deformation of the jaw line due to osteomas.

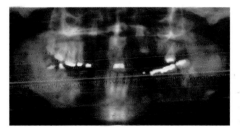

X-ray demonstrating supernumary teeth and osteoma.

Age of onset　　　6–70 years

Epidemiology　　　1 in 10,000

Inheritance　　　Autosomal dominant. Penetrance varies depending on the position of the mutation, but in most patients all symptoms (100% penetrance) are seen by the age of 40 years.

Chromosomal location　　5q21

Gene	*APC*: 15 exons, with exon 15 accounting for 77% of the coding region. Encodes APC, a 2843 amino acid protein. There is a marked genotype-phenotype correlation, with mutations at the 5' end of the gene (exons 4–5) producing an attenuated phenotype. Mutations that occur in codons 1250–1464 (exon 15) result in a severe phenotype. There are also correlations between the position of mutations and the presence or absence of CHRPE and desmoid tumors. Germline mutations are present in FAP and somatic mutations are found in sporadic tumors.
Effect of mutation	APC molecules act as dimers, affecting the binding of catenins to the cell adhesion molecule E-cadherin. Mutated APC molecules form defective dimers that result in loss of function and the onset of FAP. An explanation of the genotype-phenotype correlation is that mutations in the 5´ end of the gene result in severely truncated APC proteins that are unable to bind with the wild-type protein to affect function. However, mutations in the 3´ end result in a larger protein, which allows for heterodimer formation and blocking of APC function.
	Blocking of APC function allows the accumulation of β-catenin, which eventually results in increased transcription of genes, including *c-myc* and *cyclin D1*; both of which are involved in the regulation of the cell cycle.
Diagnosis and counselling	Patients at risk of developing FAP should be given an annual colonoscopy to screen for the presence of polyps. Screening should start at 10 years of age, or earlier if symptoms occur. If multiple polyps are present, elective colectomy is planned: either a total colectomy with ileorectal anastomosis (in this case a regular rectal surveillance would still be necessary) or panproctocolectomy, with or without a pouch. Upper GI endoscopy should be started from age 20 years in affected patients and repeated every 3 years.
	Genetic testing is now available, although original mutation screening is still a long process. Once a mutation is identified within

a family, predictive testing should be offered from the age of 10 years. This facilitates appropriate management of mutation carriers and non-carriers. Most families with two or more affected members are suitable for linkage analysis if mutation screening is unsuccessful.

Fanconi anemia

(also known as: FA; Fanconi pancytopenia)

MIM   227650

Clinical features   FA is characterized by bone marrow hypoplasia in association with developmental anomalies. The hematological presentation includes anemia, leukopenia and thrombocytopenia. Approximately 5–20% of patients die from leukemia. About 50% of patients have radial ray abnormalities ranging from bilateral absent thumbs to a unilateral hypoplastic thumb or bifid thumb. Other features include growth retardation, 'café au lait' (CAL) spots, renal malformations and gastrointestinal, cardiac, central nervous system and skeletal abnormalities. The phenotype is variable: hypoplastic anemia with no congenital abnormalities may be seen in families with the full syndrome. As well as leukemias, there is an increased risk of hepatocellular carcinoma. The mean survival time of patients following the onset of hematological symptoms is 5 years.

Genetic heterogeneity of FA has been investigated with complementation groups. There are 8 different complementation groups, FA(A–G), with one gene accounting for one clinical complementation group, except for FA-D.

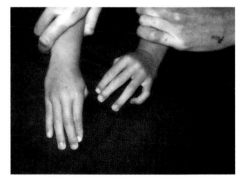

Note absence of thumbs in a child with FA.

Age of onset	Pancytopenia presents between 5–10 years of age
Epidemiology	1 in 360,000
Inheritance	Autosomal recessive
Chromosomal locations	16q24.3; *FANCA*
	9q22.3; *FANCC*
	3p25.3; *FANCD2*
	6p21.3; *FANCE*
	11p15; *FANCF*
Gene	*FANCA*: 43 exons, encodes for a 1455 amino acid protein. Accounts for 60–65% of cases.
	FANCC: 14 exons, encodes for a 558 amino acid protein. Accounts for 20% of cases.
	FANCD2: 44 exons, encodes for a 1451 amino acid protein.
	FANCE: 10 exons, encodes for a 536 amino acid protein.
	FANCF: 1 exon, encodes for a 374 amino acid protein.
	FANCG: 14 exons, encodes for a 622 amino acid protein.
Function of protein	*FANCA, C, E, F* and *G* encode for proteins that form a nuclear complex. These proteins are probably involved in a DNA-damage response pathway. It is thought that activation of this complex by DNA damage may lead to activation of *FANCD2*, which may then facilitate DNA repair.
Diagnosis and counselling	Cells from patients with FA demonstrate chromosomal fragility. These cells are hypersensitive to the alkylating agents, mitomycin C and diepoxybutane. These agents are now used as diagnostic tests. Management of FA includes regular support by transfusion and

administration of steroids and androgens to stimulate hematopoiesis. Bone marrow transplantation is the only curative treatment.

Genetic testing

Diagnosis is usually dependent on a mitomycin C test, which can also be used for prenatal diagnosis. Due to the difficulty of screening genes involved in FA, this technique is used in preference to genetic testing.

Hereditary non-polyposis colorectal cancer

(also known as: HNPCC)

MIM	120435 (*MSH2*); 120436 (*MLH1*)
Clinical features	Increased tendency to colorectal cancer at an average age of 41 years. There is an increased incidence of right-sided tumors that are frequently multiple, arising from adenomatous polyps. The number of polyps is much less than in FAP. The lifetime risk of bowel cancer in known gene mutation carriers is about 80% in men and 40–60% in women. The risk of endometrial cancer is 40–50% in women. There is also an increased risk of cancer of the ovary, stomach, small intestine, renal tract and biliary tract. Transitional cell tumors of the ureters are specific to HNPCC. The diagnostic (modified Amsterdam) criteria are: at least three relatives with an associated HNPCC, one of whom must be a first degree relative to the other two, at least two successive generations must be affected and one of the affected cases should have been diagnosed before 50 years of age. FAP must be excluded and tumor pathology should be verified.
Screening	Known gene carriers should be offered colonoscopies every 18 months from 25–30 years. Those at 50% risk, or fulfilling the Amsterdam criteria, are offered screening on a 2-yearly basis. Endometrial screening is by transvaginal ultrasound and hysteroscopy with pipelle biopsy on an annual basis. Annual urine microscopy has been advocated for screening the kidney for transitional cell carcinoma. The benefits of prophylactic colectomy are under debate.
Age of onset	25–60 years
Epidemiology	1–5% of all colorectal cancers
Inheritance	Autosomal dominant

Chromosomal location	2p22; *MSH2*
	3p21; *MLH1*
	2q31; *PMS1*
	7p22; *PMS2*
	5q; *MSH3*
	2p16; *MSH6*
	14q24.3; *MLH3*
Gene	All the above are mismatch repair genes. The majority of mutations are in the *MSH2* and *MLH1* genes. Although mutations span the entire gene sequence, two specific mutations account for 63% of HNPCC families in Finland. Only two mutations in *PMS2* and one in *PMS1* have been reported. *MSH6* mutations have been reported in HNPCC families with a greater frequency of endometrial malignancy. These families also seem to have an older age of onset of malignancies and incomplete penetrance of the mutations.
Function of protein	Errors can occur during replication of DNA within any dividing cell. Expressed proteins from mismatch repair genes recognize mismatched base pairs, excise them and replace them with the correct nucleotides. If this process is not functioning properly, the repeated DNA replication errors can result in genetic instability in which regions of DNA composed of short repeated sequences become characteristically altered. This is known as microsatellite instability (MSI). Somatic mutations may arise in unstable genes during tumorigenesis resulting in acceleration of tumor progression and an earlier onset of malignancy. The majority of mutations affect protein-protein or protein-DNA interactions.
Genetic testing	This is now available, although original mutation screening is still a long process. For families fulfilling modified Amsterdam criteria, mutation screening should be performed. Families who do not fulfill

the criteria should have tumor blocks sent for MSI testing. If the blocks show high levels of MSI, (instability at ≥ 2 out of 5 loci or at $\geq 30\text{--}40\%$ of studied loci) mutation screening of DNA should be undertaken. Once a mutation is identified within a family, predictive testing should be offered to adults to enable targeted screening.

Amsterdam criteria

Diagnostic criteria for families with HNPCC (Vasen HF et al. Diseases of the Colon and Rectum 1991;34:424–5.):

- There should be at least three relatives with histologically verified colorectal cancer

- One affected relative should be a first degree relative to the other two

- At least two successive generations should be affected

- In one of the relatives colorectal carcinoma should be diagnosed under the age of 50 years

- Familial adenomatous polyposis coli (FAP) should be excluded

These diagnostic criteria were drawn up as a research tool, and as such, not all affected individuals fulfill the diagnosis. Therefore, in 1999, modified criteria were suggested, which included extra colonic malignancies (Vasen HF et al. Gastroenterology 1999; 116:1453–56). These diagnostic criteria are now those most commonly used.

Modified Amsterdam criteria

- There should be at least three relatives with an HNPCC-associated cancer (colorectal cancer, endometrial, small bowel, ureter or renal pelvis malignancy)

- One affected relative should be a first-degree relative of the other two

- At least two successive generations should be affected

- At least one malignancy should be diagnosed before age 50 years

- FAP should be excluded in the colorectal cancer case(s)

- Tumors should be verified by pathological examination

Juvenile polyposis

(also known as: JPS; juvenile intestinal polyposis)

MIM	174900

Clinical features
This condition is characterized by multiple hamartomatous polyps—in contrast to the solitary polyps that occur in 2% of children. The polyps are distinguished from those found in Peutz–Jeghers syndrome by the predominant stroma, cystic spaces and abundant lamina propria. Presentation often includes abdominal pain, anemia, intussusception, chronic bleeding and failure to thrive. The polyps occur mainly in the colon but are also found in the small bowel and in the stomach. There is an increased risk of malignancy especially in the colon, stomach, duodenum and pancreas. The lifetime risk of GI malignancy may be as high as 60%. Juvenile polyps are also found in Peutz–Jeghers syndrome and Cowden disease.

Screening
Unaffected patients who are thought to be at risk should be offered regular colonoscopies. Multiple random biopsies of flat mucosa, as well as polyps, are advisable since carcinomas do not always originate from polyps. It is suggested that screening starts from about 10 years of age. If multiple polyps are present, making colonoscopy and multiple biopsies difficult, elective colectomy is offered from about 20 years of age. Upper GI endoscopy should be started from 20 years of age, in affected patients, on a regular basis.

Age of onset
Polyps may develop from childhood; malignancy usually begins in the third decade.

Epidemiology
Rare

Inheritance
Autosomal dominant

Chromosomal location
18q21.1; *SMAD4*
10q22–q23; *BMPR1A*

Gene	*SMAD4* (also called *DPC4*): 11 exons, germline mutations account for 8–26% of JPS. Somatic mutations have been found in a variety of other gastrointestinal malignancies.
	BMPR1A: 11 exons
Function of protein	The protein encoded by the *SMAD4* gene is involved in transmitting signals from transforming growth factor-β (TGF-β). The TGF-β superfamily is known to be important in the control of cell growth. If *SMAD4* is inactive, the cell becomes unresponsive to TGF-β, which is significant in the control of colonic epithelium. *BMPR1A* codes for a receptor that mediates bone morphogenetic protein (BMP) signaling through SMAD4, and is one of the TGF-β superfamily. It is interesting that disruption of two separate elements of BMP signaling results in the same phenotype of JPS.
Genetic testing	Mutation screening is available on a research basis only

Li–Fraumeni syndrome

(also known as: LFS)

MIM	151623

Clinical features

Classic LFS families have a proband (index case) with sarcoma, a first degree relative with any cancer and a further first or second degree relative with any cancer or any sarcoma—all of which should be diagnosed before the age of 45 years. The tumor spectrum includes early onset breast cancer, adrenocortical tumors, bone and soft tissue sarcomas, acute leukemia, melanoma, germ cell tumors, gastric and pancreatic malignancy, and lung carcinoma.

Li–Fraumeni-like syndrome (LFL) includes a proband diagnosed with any childhood tumor or sarcoma, brain tumor or adrenocortical tumor under 45, with a first or second degree relative with a typical LFS tumor at any age and another relative with any cancer under the age of 60 years. In LFS families, the risk of malignancy in gene carriers is 50% by the age of 30 years and 90% by 70 years. Of the breast cancers in these families, 79% are diagnosed before the age of 40 years. Most soft tissue sarcomas and adrenocortical carcinomas occur in the first 5 years of life, with osteosarcomas occurring in adolescence. It is common for those individuals who survive childhood tumors to develop multiple tumors in later life.

Screening

Screening at risk individuals is extremely difficult because of the wide spectrum of malignancies associated with LFS. Annual skin, general and neurological examinations are advised, with annual abdominal ultrasound during childhood and adolescence. Mammography is suggested from 30 years of age, or, where available, MRI screening of the breast from the age of 25 years. A high degree of suspicion must be used when investigating symptomatic patients.

Age of onset

From birth

Epidemiology	Rare (less than 100 families reported)
Inheritance	Autosomal dominant
Chromosomal location	17p13.1; *TP53*
	22q12.1; *CHK2*
Gene	*TP53*: 11 exons spanning 20 kb of genomic DNA, encoding p53 (a 393 amino acid protein). The majority of mutations are within exons 5–8. Mutations are found in 70% of LFS and 20% of LFL patients. Up to 80% of childhood adrenocortical tumors are due to germline mutations of *TP53*. Somatic mutations in *TP53* have been described in most tumor types.
	Mutations in *CHK2* have been described in one family with LFS and one family with LFL.
Effect of mutation	p53 is involved in cell-cycle regulation. In normal cells with damaged DNA, p53 functions at the G1 checkpoint to delay progression to S-phase allowing time for DNA repair. If the damage is too great, p53 functions to promote apoptosis. It also appears to be directly involved with DNA repair. Therefore, mutations in p53 may result in the progression of a cell through repeated divisions allowing DNA damage to accumulate. *TP53* is a 'gatekeeper' tumor suppressor gene, important in the control of the cell cycle and apoptosis. Mutations downstream of many other tumor suppressor genes seem to involve *TP53*.
Genetic testing	Mutation testing is available, however, the utility of predictive testing is under dispute due to the lack of effective screening and prevention programs. Testing in childhood is a particularly contentious issue.

Muir–Torre syndrome

(also known as: cutaneous sebaceous neoplasms and keratoacanthomas with GI and other carcinomas)

MIM	158320
Clinical features	A variant of HNPCC characterized by the association of skin lesions with GI malignancies. The skin lesions include sebaceous adenoma, epithelioma, carcinoma and basal cell epithelioma with sebaceous differentiation. Keratoacanthoma and basal cell carcinoma have also been described. There is also an increased risk of visceral malignancy, especially GI malignancy. Colon cancer develops in 50% of patients, with 15% of patients developing endometrial cancer. Ureteric cancers are also described. In about 20% of patients the skin lesions appear before the visceral malignancies.
Screening	Known gene carriers should be offered colonoscopies every 18 months from 25–30 years. Those at 50% risk are offered screening on a 2-yearly basis. Endometrial screening is by transvaginal ultrasound and hysteroscopy on an annual basis. The utility of prophylactic colectomy is under debate.
Age of onset	25–70 years
Epidemiology	About 200 cases reported
Inheritance	Autosomal dominant
Chromosomal location	2p22; *MSH2*
	3p21; *MLH1*
Gene	The majority of mutations are in the *MSH2* gene with a smaller proportion of families having mutations in *MLH1*. Both are mismatch repair genes.

Effect of mutation

MSH2 and *MLH1* gene products recognize mismatched base pairs, excise them and replace them with the correct nucleotides. If this process is not functioning properly, the repeated DNA replication errors result in genetic instability in which regions of DNA composed of short repeated sequences become characteristically altered. This is known as microsatellite instability (MSI). Somatic mutations may arise in unstable genes during tumorigenesis, resulting in acceleration of tumor progression and an earlier onset of malignancy.

Genetic testing

This is now available, although original mutation screening is still a lengthy (months) process. Once a mutation is identified within a family, predictive testing should be offered to adults to enable targeted screening.

Multiple endocrine neoplasia type 1

(also known as: MEN1)

MIM	131100
Clinical features	Glandular hyperplasia and/or tumors of the parathyroid cells, pancreatic islet cells (gastrinoma, insulinoma, glucagonoma, vasoactive intestinal polypeptidoma [VIPoma] and pancreatic polypeptidoma [PPoma]) and the anterior pituitary (prolactinoma, somatotrophinoma, corticotrophinoma, non-functioning tumor). Tumors of the adrenal cortex, carcinoid, lipomas and angiofibromas have been described. Primary hyperparathyroidism is the most common feature and occurs in 95% of patients. Gastrinomas account for 50% of pancreatic islet cell tumors and, due to severe multiple peptic ulceration, are the major cause of morbidity and mortality. (Hypergastrinemia is known as the Zollinger–Ellison syndrome.) The majority of the anterior pituitary tumors secrete prolactin (60%), with 25% secreting growth hormone, and 3% secreting adrenocorticotrophin. The rest of the tumors are non-functioning. There is intrafamilial variability in the course of the disease.
Screening	Unaffected patients who are thought to be at risk need to be screened. Current recommendations suggest biochemical screening of serum calcium and prolactin in all individuals annually. Baseline pituitary and abdominal imaging should be carried out and then repeated every 5 years. A full history and examination should be undertaken to elicit symptoms and signs of hypercalcemia, nephrolithiasis, pituitary abnormalities (including visual field loss) and cutaneous abnormalities such as lipomas. Further endocrine investigation should be done if any symptoms are present. Screening should be initiated from around 10 years of age.
Age of onset	8–81 years

Epidemiology	1 in 30,000–50,000
Inheritance	Autosomal dominant. Penetrance is at least 50% by 20 years of age and >90% by 40 years of age.
Chromosomal location	11q13
Gene	*MEN1*: 10 exons with a 1830 bp coding region. The gene encodes a 610 amino acid protein called menin. More than 10% of mutations are *de novo*. Unlike MEN2, there is no apparent genotype-phenotype correlation. Germline mutations are present in *MEN1* and somatic mutations have been detected in sporadic parathyroid tumors.
Effect of mutation	The exact role of menin in tumorigenesis is not known. Menin is localized primarily in the nucleus. It is known to interact with an activator protein-1 (Ap-1) transcription factor, JunD, suggesting a role in transcriptional regulation. Recent work has demonstrated that inactivation of menin results in loss of TGF-β-mediated cell inhibition.
Genetic testing	Theoretically available, but the lack of genotype-phenotype correlation, with a wide spread of mutation throughout the gene, means that the testing is labor-intensive. However, once mutation screening is successful in an individual family, predictive genetic testing in asymptomatic at-risk individuals will enable targeted biochemical screening.

Multiple endocrine neoplasia type 2A

(also known as: MEN2A)

MIM	171400
Clinical features	Multiple endocrine neoplasia with medullary thyroid malignancy, pheochromocytomas and parathyroid adenomas. Ninety percent of the medullary carcinomas are bilateral and multifocal, and these affect the majority of patients. Pheochromocytomas will develop in 50% of patients, of which about 70% will be bilateral, and 10% will become malignant. Hyperparathyroidism will develop in 15–30% of patients.
Screening	Unaffected relatives who are at risk of developing the disease are given basal and stimulated calcitonin, serum calcium and parathyroid level screening and annual 24 h urinary catecholamine tests from 6–35 years. Known mutation carriers are offered prophylactic thyroidectomy from 6 years, and the above screening on an annual basis.
Age of onset	6–35 years, although there is marked variability between and within families.
Epidemiology	1 in 30,000
Inheritance	Autosomal dominant. There is 93% penetrance by the age of 35 years.
Chromosomal location	10q11.2
Gene	*RET* (REarranged during Transfection) proto-oncogene: 21 exons, encodes a receptor tyrosine kinase protein (RET protein). Mutations have been found in 98% of MEN2A families. Approximately 85% of all kindreds have a mutation at codon 634.

Effect of mutation
RET protein is expressed in derivatives and tumors of neural crest origin. Receptor tyrosine kinases are involved in extracellular signaling for processes controlling cell growth, differentiation and apoptosis. This depends upon recognition of specific ligands, which initiate autophosphorylation, followed by intracellular signal transduction. The mutation in codon 634 activates the receptor tyrosine kinase, which leads to dimerization of the receptor monomer, mimicking binding of the ligand to the receptor.

Genetic testing
This is now widely available and should be used to direct management of individuals within a family. In a known mutation family, predictive testing should always be offered before any prophylactic surgery is considered.

Multiple endocrine neoplasia type 2B

(also known as: MEN2B)

MIM	162300
Clinical features	The association of medullary thyroid carcinoma, pheochromocytoma and multiple neuromas with a marfanoid habitus. An association with ganglioneuromatosis of the GI tract and medullated corneal nerve fibers has also been described. The multiple neuromas occur on the lips, the anterolateral surface of the tongue and the conjunctiva. MEN2B accounts for 5% of all MEN2. Parathyroid disease is not a feature of MEN2B. Thyroid carcinoma and the mucosal neuromata occur in nearly all patients with MEN2B, and pheochromocytomas occur in 50%. The medullary thyroid cell carcinomas are particularly aggressive, with the average age at death being about 21 years.
Screening	Unaffected relatives who are at risk of developing the disease are given basal and stimulated calcitonin level screening and annual 24 h urinary catecholamine tests from infancy to 35 years—if genetic screening is not available. If genetic testing is available, screening is targeted only to those who have tested mutation-positive. Known mutation carriers are offered prophylactic thyroidectomy from infancy as cases of thyroid cancer have been described in patients as young as 3 years.
Age of onset	Infancy, although there is marked variability between and within families. Mucosal neuromas are usually the first signs of MEN2B.
Epidemiology	Rare
Inheritance	Autosomal dominant. There is 100% penetrance by the age of 35 years.
Chromosomal location	10q11.2

Gene	*RET* (REarranged during Transfection) proto-oncogene: 21 exons, encodes a receptor tyrosine kinase protein (RET protein). Fifty percent of the mutations that cause MEN2B appear to be *de novo*. More than 95% of cases are caused by a single point mutation at codon 918 (exon 16). Approximately 4% are caused by a mutation at codon 883 (exon 15).
Effect of mutation	RET protein is expressed in derivatives and tumors of neural crest origin. Receptor tyrosine kinases are involved in extracellular signaling for processes controlling cell growth, differentiation and apoptosis. This depends upon recognition of specific ligands that initiate autophosphorylation, followed by intracellular signal transduction. The M918T mutation alters the substrate specificity so that it recognizes different ligands. It is not known how the A883F mutation alters the function of the protein.
Genetic testing	This is now widely available and should be used to direct management of individuals within a family. In a known mutation family, predictive testing should always be offered before any prophylactic surgery is considered.

Neurofibromatosis type 1

(also known as: NF1)

MIM	162200

Clinical features   Characterized by >6 café-au-lait (CAL) patches (which fade with age), peripheral neurofibromas (develop around puberty), Lisch nodules (60–80%) and axillary freckling (80%). Other features include macrocephaly (45%), short stature (30%), plexiform neurofibromas (15%) and learning difficulties (15–20% special school education, 20–30% normal school with specific learning difficulties). Severe mental retardation is rare, occurring in only 3% of patients. Complications occurring in <10% of cases include epilepsy, scoliosis, pseudoarthrosis and tumors. Optic nerve gliomas are the most common CNS tumor (10%). The overall lifetime risk of related malignancy with NF1 may be as high as 15% and includes astrocytoma, rhabdomyosarcoma, chronic myeloid leukemia, pheochromocytoma and malignant nerve sheath tumors. Vestibular schwanommas are not a feature of NF1. There is a large degree of variability both between and within families.

Screening   Patients with NF1 should be seen on an annual basis for general clinical review including an assessment of growth and development in children. Children should have an annual ophthalmic assessment until the age of 10 years to detect symptoms of optic gliomas. MRI scans should only be performed if there are signs or symptoms of a neurological deficit. The clinical assessment should also include blood pressure measurement and assessment of scoliosis. Patients at 50% risk of inheriting NF1 should have a skin examination for CAL patches, neurofibromas and axillary freckling. If CAL spots are not present by the age of 6 years, the child is unlikely to be affected. Ophthalmic examination for the presence of Lisch nodules is also suggested.

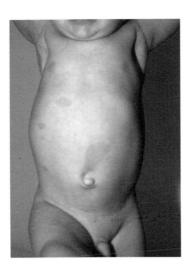

Café au lait spots in NF1.

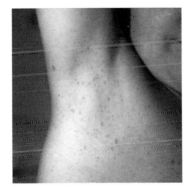

Axillary freckling in NF1.

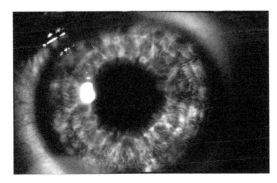

Lisch nodules in NF1.

Age of onset	Birth onwards
Epidemiology	1 in 4000 (incidence 1 in 2500)
Inheritance	Autosomal dominant
Chromosomal location	17q11.2
Gene	*NF1*: 60 exons spanning about 350 kb of genomic DNA. Four alternatively spliced transcripts have been identified. The gene encodes a 250 kDa protein (neurofibromin). Mutations are *de novo* in 50% of cases.
Effect of mutation	Neurofibromin appears to act as a tumor suppressor by down regulating p21 (ras). p21 has a major influence on cell growth and regulation. It has been postulated that mutations in neurofibromin will result in abnormal RAS signaling pathways, thereby contributing to tumor development.
Genetic testing	Mutation testing is difficult due to the high new mutation rate and the size of the *NF1* gene. Therefore, in clinical practice mutation testing and predictive testing are rarely undertaken. Usually, it is easy to make a diagnosis in the clinic.

Neurofibromatosis type 2

(also known as: NF2)

MIM   101000

Clinical features   The average age of presentation of NF2 is 22 years, and 85% of the cases present with bilateral vestibular schwannomas (acoustic neuromas). Other diagnostic criteria are a family history of NF2 plus a unilateral vestibular schwannoma (6%), or any two of meningioma (45%), glioma, posterior subcapsular lenticular opacity (60%), skin neurofibroma or schwannoma of any peripheral or cranial nerve. Café-au-lait spots are not as common in NF2 as they are in NF1. As with NF1 there is a great deal of heterogeneity in the manifestations of the disease although there does appear to be a correlation of symptoms within a family. About 6% of NF2 patients develop a malignancy. Ninety percent of patients become deaf due to vestibular schwannomas. Individuals with NF2 should be managed by a multidisciplinary team, including geneticists, neurosurgeons and otolaryngologists.

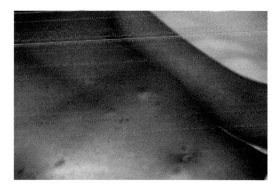

Cutaneous schwannomas in NF2.

Screening   Those individuals at risk of developing NF2 should be seen on an annual basis, with a detailed ophthalmic examination for cataracts from birth. From 2–10 years of age, assessments should include history, neurological and skin examination. From 10 years onwards, audiological testing including brain stem evoked responses is added to the examination. MRI scanning is carried out at the age of 14 years and every 3 years afterwards. A normal MRI scan at 30 years suggests that inheritance of NF2 is unlikely.

Age of onset   Symptomatic before 10 years in 10% of cases, all patients will develop a CNS tumor by the age of 55 years.

Epidemiology   1 in 33,000

Inheritance   Autosomal dominant

Chromosomal location   22q12.2

Gene   *NF2*: 17 exons, encodes a 595 amino acid protein. The protein is called merlin or schwannomin. Fifty percent of mutations are *de novo*. Truncating mutations appear to be associated with severe, multi-tumor, early-onset disease.

Effect of mutation The NF2 protein is related to a family of proteins that connect the cytoskeleton to membrane proteins. Experiments expressing mutant merlin proteins in mammalian cells demonstrated significantly altered cell adhesion. The influence of merlin on cell growth may be dependent on specific cytoskeletal rearrangements.

Genetic testing   Mutation testing is available and subsequent predictive testing will assist in directing screening.

Nijmegen breakage syndrome

(also known as: NBS; ataxia-telangiectasia variant V1; microcephaly with normal intelligence, immunodeficiency and lymphoreticular malignancies)

MIM	251260
Clinical features	Characterized by microcephaly, growth retardation and characteristic craniofacial features that become more obvious with age. The midface is prominent with a long nose, receding forehead, receding mandible, upslanting palpebral fissure and large ears with dysplastic helices. There is no evidence of ataxia. Mental development is normal in 35% of cases, 45% have mild learning difficulties and 25% have moderate learning difficulties. Pigmentary abnormalities are usual. Immunodeficiency results in repeated infections and, along with radiation hypersensitivity, contributes to the increased risk of malignancy. The highest risk is of lymphoreticular malignancies, although medulloblastoma, glioma, neuroblastoma and rhabdomyosarcoma have been described.
Screening	Screening in NBS is difficult although a regular full blood count (FBC) and clinical examination for lymphadenopathy would be useful. Due to radiation sensitivity, radiotherapy should be avoided in the treatment of malignancies as it may induce further tumors. Cytostatic agents should be the treatment of choice, with radiomimetics such as bleomycin being avoided and the dose of chemotherapy being reduced.
Age of onset	The microcephaly and dysmorphic features are noted from birth onwards, malignancies develop from mid-childhood.
Epidemiology	Rare, fewer than 100 patients reported.
Inheritance	Autosomal recessive
Chromosomal location	8q21

Gene	*NBS1*: 16 exons spanning 50 kb, encodes a 750 amino acid protein called p95 or nibrin.
Effect of mutation	Nibrin/p95 is a component of the MRE11/RAD50 double-strand break repair complex. This protein is absent in the cells of NBS patients and the complex does not form around radiation-induced foci. The protein also interacts with ataxia-telangiectasia mutated (ATM) in pathways involving the MRE11/RAD50 complex, the S-phase checkpoint and the rescue of hypersensitivity to radiation.
Genetic testing	Mutation screening is available on a research basis only

Peutz–Jeghers syndrome

(also known as: PJS)

MIM   175200

Clinical features   Characterized by hamartomatous GI polyps and mucocutaneous pigmentation. The polyps can be found anywhere in the GI tract, but 70–90% of patients have polyps within the small bowel. Complications typically occur within the first decade and include intussusception, small bowel obstruction, rectal bleeding and volvulus. Pigmentation in the form of freckles on the lips, buccal mucosa, fingers, toes and vulva usually occurs. There is an increased incidence of malignancy of the following sites: GI tract, breast, pancreas, ovary and gall bladder. PJS patients have less chance of surviving these cancers than the general population.

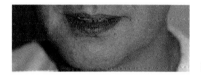

Pigmentation of lips seen in PJS.

Screening   Some suggest that an upper GI endoscopy and small bowel meal, with follow-up, should be undertaken every 2 years from the age of 10 years, 3-yearly colonoscopy and annual transvaginal ultrasound from the age of 25 years, and annual mammography from the age of 35 years. The difficulty of small bowel imaging and the increased radiation exposure from the procedure, with the preponderance of GI malignancies in the stomach or large bowel, would suggest that small bowel screening should not be undertaken so frequently. Further work is needed to clarify the exact risks of malignancy in PJS and the value of such an intensive screening program.

Age of onset   From birth onwards. The mucocutaneous changes tend to occur in the first decade and fade from the third decade onwards.

Epidemiology	Rare
Inheritance	Autosomal dominant
Chromosomal location	19p13.3
Gene	*STK11* (also known as *LKB1*): 9 exons spanning 23 kb, encodes a 443 amino acid protein (a serine/threonine protein kinase). Germline mutations have been identified in 60% of familial and 50% of sporadic cases.
Function of protein	The function of the protein has not been elucidated. In the mouse model, it has been demonstrated that Lkb1 is probably a nuclear protein, with a nuclear localization signal within the sequence. *STK11* physically associates with p53 to regulate specific apoptotic pathways. It is hypothesized that if *STK11* is inactive, disruption of this pathway results in absence of apoptosis which may then result in polyp formation.
Genetic testing	Mutation screening is available on a research basis only

Retinoblastoma

(also known as: RB)

MIM	180200
Clinical features	Retinoblastomas arise from primitive retinal cells. Ninety percent of cases present before the age of 5 years. Thirty percent of cases are bilateral and have a younger age of onset (mean age is 8 months vs. 25 months), and only 10% of cases have a positive family history. Only 3% of patients with a germline mutation go on to develop a pinealoma, or a suprasellar tumor. The risk of osteogenic sarcoma is greatly increased in germline mutation cases and appears to be related to previous radiotherapy. An increased frequency of malformations has also been reported.
Screening	Children at risk (siblings or offspring of retinoblastoma patients) should have a retinal examination at birth and monthly until 3 months, then examinations with general anesthesia every 3 months until 2 years, then every 4–5 months until 3 years. Biannual examinations can then take place without anesthesia until 5 years and then annually until 11 years of age.
Age of onset	Birth to 11 years
Epidemiology	1 in 20,000 (familial 1 in 50,000)
Inheritance	Autosomal dominant. Families with incomplete penetrance have been reported.
Chromosomal location	13q14.1–14.2
Gene	*RB1*: 27 exons, encodes a 928 amino acid protein. No genotype/phenotype correlation has been reported.

Effect of mutation   The Rb protein is involved in both the regulation of cell proliferation and apoptosis. Rb interacts with E2F transcription factor which is involved in controlling the transition of a cell from G_1 to S phase. This pathway also induces p53-dependent cell death by the deregulation of E2F1. Deregulation of this pathway (which also involves p16^{INK4a}, D cyclin/cdk4 kinase, p19^{ARF} and p73) results in a number of different malignancies.

Genetic testing   Mutation screening is available, although it is a long process. Predictive testing can be used to determine whether ophthalmological screening of a neonate is necessary.

Sotos syndrome

(also known as: cerebral gigantism)

MIM   117550

Clinical features   Characterized by excessively rapid growth and distinctive facial
features including frontal bossing, high hair line with sparse hair,
prominent jaw, downslanting palpebral fissures, a long pointed chin
and a high arched palate. Birth length and weight are abnormally
large, with most cases having a height above the 97th percentile
throughout childhood and adolescence. The final height is frequently
below this percentile. Most patients are macrocephalic with
advanced bone age. There is a variable pattern of developmental
delay. It has been suggested that there is a risk of benign or
malignant tumors (4%) including hepatoblastoma, Wilms tumor,
neuroblastoma, gastric carcinoma, and leukemias.

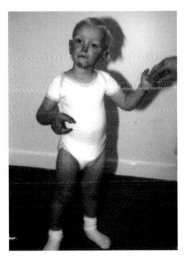

Sotos syndrome demonstrating
general overgrowth.

Note the sparse hair, long pointed chin and frontal bossing.

Screening

Screening for malignancy in Sotos syndrome is controversial. As with Beckwith–Wiedemann syndrome, it has been suggested that ultrasound examination of the abdomen should initially be undertaken on a quarterly and then biannual basis during childhood to exclude hepatoblastoma and Wilms tumor. In view of the slightly increased risk of gastric carcinoma in adults, it would seem reasonable to offer *Helicobacter* screening and eradication therapy from the age of 20 years.

Age of onset

The features of overgrowth are present both pre- and post-natally

Epidemiology

200–250 cases described

Inheritance

Autosomal dominant

Gene

Not known

Effect of mutation

Not known

Sotos syndrome

Tuberous sclerosis

(also known as: TS; Bourneville disease)

MIM 191100 (*TSC1*)

191092 (*TSC2*)

Clinical features An association of a number of features including facial angiofibromas (adenoma sebaceum) in 80–90% of patients, from 1–100 hypomelanotic patches (80–90%), shagreen patches (20–40%), forehead patches (25%), ungual fibromas (20–50%) and dental enamel pits (48%). The syndrome is so-called due to the cortical tubers present in 95% of patients, best demonstrated on brain MRI. Epilepsy occurs in 60% of patients, and leads to learning disability in 40–60%. The severity of the learning disability is related to the severity of the epilepsy. Cardiac rhabdomyomas occur in 50% of children and regress with age. Renal angiomyolipomas are found in 67% of cases with an increased incidence of renal cysts and renal cell carcinoma (3%). Retinal hamartomas and depigmentation are also described but are generally asymptomatic.

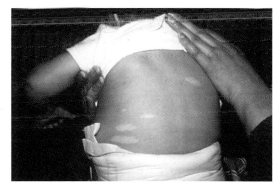

Hypomelanotic patches on the back of a child with TS.

Screening	Affected patients should be reviewed annually with a general examination and systemic inquiry. Renal ultrasound is controversial, but in patients with known renal disease it is suggested every 1–2 years. Pre- and neonatal fetal echocardiogram should be done to exclude cardiac rhabdomyomas. At-risk relatives should undergo examination, looking for cutaneous manifestations and dental pits. An ophthalmic opinion should be sought and a brain MRI requested.
Age of onset	Hypomelanotic patches from birth, facial angiofibromas from puberty, ungual fibromas and shagreen patches may develop in adults. Epilepsy tends to occur in the first years of life; only 4% of TS patients present with epilepsy in adulthood.
Epidemiology	1 in 5800
Inheritance	Autosomal dominant
Chromosomal location	9q34 (*TSC1*) 16p13.3 (*TSC2*)
Gene	*TSC1*: 23 exons, encodes a 1164 amino acid coiled-coil domain protein called hamartin. Mutations in this gene account for 40–50% of mutations in TS. Learning disabilities are less common with mutations in *TSC1*. *TSC2*: 41 exons, encodes a 1807 amino acid protein called tuberin. This form accounts for 50–60% of mutations in TS Mutations in *TSC2* are more common in sporadic cases (32%) than in familial cases (9%).
Effect of mutation	Hamartin and tuberin are known to interact, this association is mediated by the coiled-coil domains of the hamartin protein. This interaction suggests that they function within the same complex, which explains their similar clinical phenotypes. Hamartin is a negative regulator of cell proliferation. Tuberin and hamartin interact

with the cyclin-dependent kinase, CDK1, suggesting that tuberin and hamartin have some affect on the activity of CDK1. CDK1 is known to trigger the G2/M transition of the cell cycle. Disruption of tuberin and hamartin may then result in aberrant regulation of the cell cycle.

Genetic testing

Mutation testing is now available. Clinical information is used to direct screening initially to either *TSC1* or *TSC2*—complete screening of both genes is time-consuming and expensive. Diagnostic and prenatal testing is available if a family mutation is detected.

von Hippel–Lindau disease

(also known as: VHL)

MIM

193300

Clinical features

Characterized by retinal angiomas (60% of patients), cerebellar hemangioblastomas (60%) and renal cell carcinomas (28%). Spinal and brainstem hemangioblastomas, pheochromocytomas, pancreatic and renal cysts have all been described. The chances of developing the following diseases before the age of 60 years are: hemangioblastoma 84%, hemangioblastomas (84%), renal cell carcinoma (70%) and retinal angiomas (70%). The diagnostic criteria include multiple retinal angiomas or multiple cerebellar hemangioblastomas, a single cerebellar hemangioblastoma or retinal angioma with a renal cell carcinoma, or a single tumor with a family history of VHL. The mean age of diagnosis of VHL is 26 years, with a median survival of 49 years.

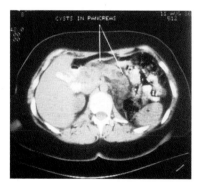

Pancreatic cysts in a patient with VHL.

Screening

Unaffected relatives at risk of developing the disease are offered screening including an annual physical examination for neurological signs and abdominal masses. Renal ultrasound, ophthalmic examination and 24 h urinary vanillylmandelic acid (VMA) tests are done on an annual basis. MRI of the head and spine is carried out every 3 years or sooner if signs or symptoms are present. The

	management of patients should take place in a multidisciplinary setting to ensure that affected patients are offered appropriate screening.
Age of onset	Infancy (rarely), usually in the teens to early 20s. There is 52% penetrance by 25 years, 80% penetrance by 35 years, 90% penetrance by 45 years, 99% penetrance by 65 years.
Epidemiology	1 in 85,000
Inheritance	Autosomal dominant
Chromosomal location	3p25
Gene	*VHL*: three exons encode a 213 amino acid protein. There appears to be a genotype/phenotype correlation with missense mutations predisposing to pheochromocytomas (59% at 50 years). Large deletions or truncating mutations are not associated with pheochromocytomas (9% at 50 years).
Effect of mutation	The VHL protein (pVHL) is involved in the regulation of hypoxia-induced vascular endothelial growth factor (VEGF) and glucose transporter-1 (GLUT-1). It is also involved in the degradation of hypoxia-inducible factor-1 (HIF-1). HIF-1 usually controls factors promoting the formation of blood vessels. If pVHL is abnormal, the degradation of HIF-1 does not take place, resulting in abnormal proliferation of blood vessels. This could account for the vascular tumors seen in VHL. Further work is ongoing to fully understand the interactions of pVHL with other proteins and to understand the genotype/phenotype correlation.
Genetic testing	This is now available and should be used to direct management of individuals within a family. A negative predictive test would remove patients from a screening program, allowing those patients at very high risk to be targeted.

Xeroderma pigmentosum

(also known as: XP)

MIM   278700 (XP-group A)

Clinical features   Xeroderma pigmentosum is a heterogeneous group of disorders with the same clinical features. Variations in the intensity of clinical aspects are described according to the gene involved (there are seven known complementation groups, A–G). These disorders are characterized by sensitivity to UV light and the subsequent development of dermatological malignancies at a young age. Within the first few years of life, freckle-like lesions occur in exposed areas. These are followed by telangiectasia and atrophy with hyperpigmentation. The risk of malignant melanoma, basal cell carcinoma and squamous cell carcinoma is increased 2000-fold. The incidence of malignant lesions of the mucous membranes, tongue and oropharynx is greatly increased. Associated neurological features include progressive mental deficiency, sensorineural deafness, cranial nerve damage, ataxia, choreoathetosis, spasticity and lower motor neuron damage. The severity of the neurological features varies, and they occur in 20% of patients, most of them have the group D disorder. Most XP cases are in the A and C complementation groups.

Management   Fibroblasts from XP patients are unable to repair UV-induced radiation damage. Therefore, children with this condition should be protected from sunlight. Careful surveillance of the skin and early treatment of any malignancies are essential.

Age of onset   Onset is before 18 months in 50% of cases. The median age of skin malignancy development is 8 years.

Epidemiology   1 in 70,000

Inheritance	Autosomal recessive
Chromosomal location	9q22.3 (*XPA*)
	2q21 (*XPB*)
	3p25 (*XPC*)
	19q13.2–q13.3 (*XPD*)
	11p12–p11 (*XPE*)
	16p13.3–p13.13 (*XPF*)
	13q33 (*XPG*)
Gene	*XPA*: encodes a hydrophilic protein involved in recruiting incision complexes to the site of DNA damage.
	XPB: (also called *ERCC3*) encodes for a protein which is a component of transcription factor TFIIH.
	XPC: encodes an 823 amino acid protein that acts as a DNA damage detector in nuclear excision repair (NER).
	XPD (also called *ERCC2*): encodes a DNA helicase involved in nucleotide excision repair of UV-induced DNA damage.
	XPE (also called *DDB2*): encodes a DNA binding protein.
	XPF (also called *ERCC4*): encodes a DNA endonuclease.
	XPG (also called *ERCC5*): encodes a DNA endonuclease.

Effect of mutation   NER involves the repair of any type of DNA damage by recognition of the lesion, removal of the damaged length by a DNA helicase, followed by synthesis and replacement of the new strand by DNA polymerases and DNA ligases. The proteins coded by the XP genes are all involved in this pathway, therefore, mutations within the gene lead to an accumulation of DNA mutations and instability.

Genetic testing   Mutation testing in not available as a service. Prenatal diagnosis is available by measuring levels of DNA repair in amniocytes exposed to UV radiation.

Xeroderma pigmentosum; variant type

(also known as: XPV)

MIM	278750
Clinical features	With the exception of neurological symptoms, XPV has the same clinical features as XP. It is characterized by sensitivity to UV light and the subsequent development of dermatological malignancies at a young age. Within the first few years of life, freckle-like lesions occur in exposed areas. This is followed by telangiectasia and atrophy, with hyperpigmentation. The risk of malignant melanoma, basal cell carcinoma and squamous cell carcinoma is increased 2000-fold. The incidence of malignant lesions of the mucous membranes, tongue and oropharynx is greatly increased.
Management	Whilst fibroblasts show marked sensitivity to the mutagenic affect of UV radiation, they have almost normal sensitivity to cytotoxic affects, unlike classical XP where there is increased sensitivity to both. Children with this condition need to be protected from sunlight. Careful surveillance of the skin is essential with early treatment of any malignancies.
Age of onset	In 50% of cases onset is before 18 months. The median age of onset is 8 years.
Epidemiology	Only a handful of cases have been described
Inheritance	Autosomal recessive
Chromosomal location	6p21.1–p12
Gene	*XPV*: encodes a 713 amino acid protein called POL H.

Function of protein   POL H is a DNA polymerase that continues replication of DNA by bypassing UV-induced dimers. Further studies are needed to elucidate the function of this protein.

Genetic testing   Mutation testing is not available as a service

2. Common cancers

BREAST CANCER

Risk assessment

Breast cancer is common, with 1 in 12 women developing the disease. The population risk in men is 1 in 1200. A family history of breast cancer is a major risk factor in women. The risk is proportional to the number of relatives affected and is inversely related to the age at diagnosis of breast cancer within the family. Hormonal factors are also known to influence the risk of breast cancer, with prolonged exposure to either endogenous or exogenous estrogens increasing the risk.

About 5–10% of breast cancer is inherited. Assessing a woman's risk of breast cancer is straightforward in a family with an autosomal dominant pattern of cancers. First degree relatives of affected women in these families have a 50% risk of inheriting a mutation and the associated risk of developing malignancy. However, most of the families that attend counselling will not have such a strong history. A large epidemiological study (Cancer and Steroid Hormone Study) in the US has provided useful information that allows the estimation of lifetime risk depending on the number of affected cases and the age of diagnosis, (Claus EB, Risch N, Thompson WD. Autosomal dominant inheritance of early-onset breast cancer. Implications for risk prediction. Cancer 1994;73:643–51). For example, a woman with a first degree relative diagnosed with breast cancer under the age of 40 years has a lifetime risk of breast cancer of 1 in 6 (double the population risk), whilst a woman whose mother and a maternal aunt were both diagnosed with breast cancer at the age of 50 years, has a lifetime risk of 1 in 3–4.

The Claus data are widely used in the UK; other countries use different models of risk assessment including the Gail model. This model incorporates family history, estrogen status, and biopsy results based on data from the Breast Cancer Detection Demonstration Project. This information has been used to produce graphs for the estimation of an individual's risk of breast cancer (Benichou et al. J Clin Oncol 1996;14:103–10). A number of computer programmes are also available for risk estimation, based on both the Claus and Gail model and are widely used.

Current recommendations for management of women with a family history of breast cancer are that women at low (actually average) risk (1 in 7–12) should be reassured, those at moderate risk (1 in 4–6) should be managed in a surgical unit, and those at high risk (1 in 3 and higher) should be referred to the regional genetics department.

Screening

Screening for breast cancer in young women has not been shown to be effective and its utility is under debate. However, preliminary evidence from small studies of high risk women and from Swedish trials (women in the general population over the age of 40 years were screened) suggests that it will be effective. Currently, the only widely available screening modality is mammography. A UK national trial assessing the use of MRI as a screening modality in very high risk women is underway, but results will not be available for a number of years.

Current recommendations for screening in the UK are that all women between the ages of 50–64 should have a mammogram once every 3 years. Women with a lifetime risk of 1 in 6 or higher should be offered an annual mammogram between the ages of 35–50 years (or 5 years younger than the youngest age at diagnosis in the family, with the minimum age at screening of 30 years). Women with a lifetime risk of 1 in 3 or higher should be offered a mammogram every 18 months from the age of 50 years.

The National Cancer Advisory Board in the USA suggests that every women over the age of 50 years should have a mammogram every 1–2 years. Any woman with an average risk of breast cancer should have a mammogram every 2 years between the ages of 40–50 years, and any woman with a family history should discuss this with her medical practitioner.

Age of onset

Familial breast cancers tend to occur before the menopause

Gene   *BRCA1*

BRCA2

TP53 (see Li–Fraumeni syndrome)

PTEN (see Cowden disease)

ATM (see ataxia telangiectasia)

The pattern of malignancies within a family may be an indication of the gene involved (see Table 1). *BRCA1* and *BRCA2* account for the majority of familial high penetrance breast cancers, but research is currently aimed at localizing *BRCA3*.

Table 1. Proportion of *BRCA1* or *BRCA2* mutations in families fulfilling the breast cancer linkage consortium (BCLC). (Under the BCLC criteria, a family with breast cancer is defined as a family with four or more cases of breast and/or ovarian cancer diagnosed under the age of 60 on the same side of the family.)

	BRCA1	*BRCA2*
all families (BCLC criteria)	50%	35%
1 epithelial ovarian cancer	80%	15%
2 or more epithelial ovarian cancers	90%	8%
4–5 breast cancers (no ovarian)	32%	10%
6 or more cases of breast cancer only	19%	66%
male breast cancer families	20%	76%

BRCA1-related cancers

(also known as: breast cancer 1 gene-related cancers)

MIM   113705

Clinical features   Early studies of breast cancer families (for fulfillment criteria see
                      Table 1) show that mutations in *BRCA1* are associated with an 85%
                      lifetime risk of female breast cancer, 30–60% lifetime risk of ovarian
                      cancer (includes malignancy of the fallopian tubes), 6–8% lifetime
                      risk of bowel cancer and increased risks of prostate cancer. The risk
                      of contralateral breast cancer after a primary breast malignancy is
                      approximately 60%. The pattern of age-specific incidence of breast
                      cancers is very different from that in the general population, with the
                      relative risks being much higher in the under 40s than in the over
                      60s. The penetrance of mutations may vary between populations,
                      with some evidence of lower penetrance mutations in the Ashkenazi
                      population, for example.

Screening and   Affected women should continue having mammograms on the
management   contralateral breast and annual transvaginal ultrasound scans of the
                      ovaries in conjunction with cancer antigen 125 (CA125) estimation.
                      Mutation carriers may be offered a single colonoscopy at the age of
                      45–50 years. Women at 50% risk of inheriting the mutation should
                      have mammographic screening as described on the previous page.
                      Ovarian screening should also be offered. Prophylactic surgery,
                      either mastectomy and/or oophorectomy, is an option for women at
                      high risk of malignancy and has been demonstrated to decrease risk
                      of cancer. Prophylactic mastectomy needs careful multidisciplinary
                      counselling, including psychological assessment. CA125
                      estimations should still be performed annually following prophylactic
                      oophorectomy in *BRCA1* carriers.

                      Given the high risk of contralateral breast cancer in *BRCA1* mutation
                      carriers, there is a good argument for altering management of initial
                      early breast cancers. In the general population, breast-conserving

surgery may be offered to a young woman. However, it is felt by some that a *BRCA1* mutation carrier should immediately be offered bilateral mastectomy. Oophorectomy is a further treatment option that should be considered.

Age of onset Breast cancer: ~35 years onwards; ovarian cancer: ~50 years onwards

Inheritance Autosomal dominant

Chromosomal location 17q21

Gene *BRCA1*: 24 exons (22 coding) encoding a 1863 amino acid protein. Mutations are found throughout the gene but some founder mutations exist. Two mutations account for the majority of inherited breast cancer within the Ashkenazi Jewish population: 185delAG (1%) and 5382insC (0.2%), along with a founder mutation in *BRCA2*. There is no strong genotype-phenotype correlation. Mutations in *BRCA1* have not been found in sporadic breast cancers.

Function of protein The BRCA1 protein has a ring finger domain that is involved in mediating protein-protein or protein-DNA interactions. It is a nuclear protein and is preferentially expressed during late G-phase and early S-phase of the cell cycle. BRCA1 appears to act at the G1–S-phase transition to arrest cell cycle progression. BRCA1 appears to be involved in the control of mitotic spindles and segregation of daughter chromosomes. It is also involved in DNA repair in conjunction with BRCA2.

Genetic testing Mutation testing is available but time consuming. Some laboratories offer a 'quick screen' looking for mutations in exons 11, 2 and 20 only. Once a mutation is identified within a family, predictive testing becomes an option.

BRCA2-related cancers

(also known as: breast cancer 2 gene-related cancers)

MIM	600185

Clinical features

Early studies of breast cancer families (for fulfillment criteria see **Table 1**) show that mutations in *BRCA2* are associated with 80% lifetime risk of female breast cancer, 10% lifetime risk of male breast cancer, 20% lifetime risk of ovarian cancer (includes malignancy of the fallopian tubes) and 14% lifetime risk of prostate cancer. There is also an increased risk of pancreatic, gallbladder and laryngeal malignancies. The risk of contralateral breast cancer after a primary breast malignancy is approximately 60%. The pattern of age-specific incidence of breast cancers is very different from those in the general population, with the relative risks being much higher in the under 40s than in the over 60s. The penetrance of mutations may vary between populations and is lower than that of *BRCA1* mutations. The age of onset of malignancy may also be slightly older.

Screening and management

Affected women should continue having mammograms on the contralateral breast and annual transvaginal ultrasound scans of the ovaries in conjunction with CA125 estimation. There is controversy over the value of prostate-specific antigen (PSA) estimation in men with a *BRCA2* mutation. It is recommended that men with *BRCA2* mutations should undergo regular breast examination. Women at 50% risk of inheriting the mutation should have mammographic screening as described previously. Ovarian screening should also be offered, although if the family mutation is known to be outside the ovarian cluster region, the risks may not be substantially increased. Prophylactic surgery, either mastectomy and/or oophorectomy are options available to women who are either proven mutation carriers or at high risk of malignancy. Prophylactic mastectomy needs careful, multidisciplinary counselling including psychological assessment.

In some regions, bilateral mastectomy is offered at initial diagnosis of an early onset breast cancer in mutation carriers due to the increased risk of contralateral breast cancer. Oophorectomy may be considered as a further treatment option.

Age of onset   Breast cancer: ~35 years onwards; ovarian cancer: ~50 years onwards

Inheritance   Autosomal dominant

Chromosomal location   13q12

Gene   *BRCA2*: 27 exons encoding a 3418 amino acid protein. Mutations are found throughout the gene but some founder mutations exist. A single mutation, 6174delT, accounts for the majority of inherited breast cancer within the Ashkenazi Jewish population along with the founder mutations in *BRCA1*. There is a weak genotype-phenotype correlation, with mutations in exon 11 between nucleotides 3035 and 6629 (the ovarian cluster region) conferring a risk of ovarian cancer of 20% as compared to mutations outside this region conferring a risk of 10%. Mutations have not been found in sporadic breast cancers.

Function of protein   BRCA2 is a histone acetyl transferase that may be involved with regulation of transcription and tumor suppressor function. It also interacts with Rad51, a protein involved in DNA repair. Along with BRCA1 it is preferentially expressed during late G-phase and early S-phase of the cell cycle. Both BRCA1 and BRCA2 are involved in pathways that may trigger p53 activity.

Genetic testing   Mutation testing is available but time consuming. Some laboratories offer a 'quick screen' looking for mutations in exons 10 and 11 only. Once a mutation is identified within a family, predictive testing becomes an option.

CENTRAL NERVOUS SYSTEM TUMORS

Risk assessment

Tumors of the central nervous system (CNS) affect between 6–13:100,000 people per year. Tumors of the CNS can occur at all ages, although most childhood tumors (mainly medulloblastomas, optic gliomas and ependymomas) arise in the posterior fossa. In contrast, most adult tumors (mainly meningiomas and gliomas) are supratentorial. Although the etiology of brain tumors is not fully understood some genetic syndromes are associated with malignancies in the CNS.

Primitive neural ectodermal tumors (PNET)

This classification includes medulloblastomas, choroid plexus carcinomas, malignant rhabdoid tumors and pineoblastomas. Medulloblastomas are the most common, accounting for 25% of all childhood brain tumors. Familial clustering has been reported. PNET tumors are associated with Gorlin syndrome, familial adenomatous polyposis, Li–Fraumeni syndrome and ataxia telangiectasia.

Glioma

Gliomas include glioblastomas and astrocytomas. Site-specific familial glioblastoma is rare but has been described. A recent study of astrocytomas reported familial aggregation in a Swedish population. These families had an increased incidence of astrocytomas but the occurrence of other brain tumors was normal. Gliomas are found in neurofibromatosis types 1 and 2, Li–Fraumeni syndrome, tuberous sclerosis, Turcot syndrome and Gorlin syndrome.

Meningioma

Meningiomas account for 15% of all brain tumors and are the most frequent form of benign tumor. Familial cases of meningioma have been described. The majority of meningiomas have abnormalities on chromosome 22; these mutations may also play a role in neurofibromatosis type 2 (NF2). Many of the familial forms of meningioma described have been shown to be NF2, and meningioma also occurs in Gorlin syndrome.

Turcot syndrome	The existence of this syndrome is controversial. Turcot originally described a family with two siblings suffering from adenomatous polyposis of the bowel and brain tumors. This was considered to be an autosomal recessive disorder. However, subsequent families with polyposis and brain tumors have been described with a clearly autosomal dominant pedigree. These families are thought to have a phenotypic variation of familial adenomatous polyposis (FAP), and indeed mutations in the *APC* gene have been described in these families. However, it is possible that there are two brain-polyposis syndromes. The first syndrome includes patients with gliomas and colorectal adenomas. The gliomas occur under the age of 20 years. The tumors in these patients show DNA replication errors in keeping with hereditary non-polyposis colorectal cancer (HNPCC), and mutations of the DNA mismatch repair genes have been demonstrated in some families. The second syndrome is FAP with CNS tumors. Further work is required to determine whether or not Turcot syndrome is a distinct clinical entity.
Gene	*APC* (see FAP)
	p53 (see Li–Fraumeni syndrome)
	PTCH (see Gorlin syndrome)
	hSNF5 (see posterior fossa brain tumors)
	NF1 and *NF2*
	TSC1 and *TSC2*
	hMLH1 and *hMSH2* (see Turcot syndrome and HNPCC)

hSNF5

This gene has been implicated in the development of malignant rhabdoid tumors of the CNS. Most cases described have had germline mutations in *hSNF5*, but as no mutations have been found in their parents this indicates that these are new mutations. However, recently, a two-generation family has been described with multiple posterior fossa tumors, in whom multiple members of the family were shown to have germline mutations. In this family, the penetrance of brain tumors was not complete. *hSNF5* is localized on 22q11.2 and encodes for a protein that is involved in chromatin remodeling, which leads to transcription regulation, which then regulates the transcription. How this relates to the development of tumors is unknown.

COLORECTAL CANCER

Risk assessment

Colorectal cancer (CRC) is a common malignancy in the developed world, with an annual incidence of 32 in 100,000 in the UK. CRCs are uncommon within Africa, Asia and South America, which suggests that dietary factors are involved in the etiology of the disease.

Cancers of the colon are more common than rectal carcinomas and, within the general population, are mostly left-sided. The presence of a right-sided tumor is suggestive of an inherited component to the disease.

Most carcinomas appear to develop from adenomas. About 50% of the general population have adenomatous polyps by the age of 70 years.

Approximately 10–20% of CRC cases are familial. In these cases there is either a strong family history of familial adenomatous polyposis (FAP) or hereditary non-polyposis colorectal cancer (HNPCC), or a few relatives will have had some form of CRC.

While risk assessment is straightforward in those families where the pattern of malignancy fulfills the diagnosis for an inherited cancer syndrome, it is less so in the majority of cases.

Epidemiological data suggest that an individual's lifetime risk of developing CRC increases from the population average of 1 in 30 to about 1 in 17 if a first degree relative has the disease. This increases to 1 in 10 if the relative is diagnosed before the age of 45 years. If two first degree relatives from the same side of the family are affected, this risk increases to 1 in 6; if three or more are affected, an autosomal dominant inheritance is probable. The recommendations regarding screening are then based on the lifetime risk of bowel cancer.

Screening

Colonoscopies are offered every 3–5 years from the age of 30 years to patients who have ≥1 in 10 risk of developing CRC. Colonoscopies are the screening modality of choice due to the preponderance of right-sided tumors in the inherited forms of CRC.

The cut-off for regular colonoscopies is a 10% lifetime risk of CRC due to the potential for perforation (1 in 300 colonoscopies).

Patients who have a first-degree relative with CRC may be offered a single flexible sigmoidoscopy around the age of 45 years with regular monitoring for fecal occult bloods, although the efficacy of this screening has yet to be proven. The screening frequency is increased if there is evidence of an inherited cancer syndrome (see FAP and HNPCC). American guidelines are more aggressive, suggesting that anyone with a first degree relative with CRC should have flexible sigmoidoscopy every 5 years from the age of 40 years.

Treatment

Trials are underway assessing the role of chemopreventive agents in CRC. These are aimed at the prevention of the development of adenomatous polyps and their subsequent progression to carcinoma. The chemopreventive agents include aspirin, other non-steroidal agents, vitamins, antioxidants and resistant starch.

Age of onset

The age at onset of bowel cancer varies. There are reports of FAP in teenagers, however, in other familial forms CRC generally occurs after the age of 25 years.

Inheritance

Autosomal dominant

Gene

APC (see FAP)

Mismatch repair genes (see HNPCC)

BRCA1/BRCA2 (both indicate an increased risk of bowel cancer in carriers of the genes)

p53 (see Li–Fraumeni syndrome)

A specific mutation in the *APC* gene, *APC I1307K*, has been found in 6% of the Ashkenazi Jewish population and is thought to account for familial aggregations of CRC that do not fulfill criteria for either FAP or HNPCC. Carriers of this mutation are thought to have a two-fold increase in the risk of developing CRC, although more work is needed to clarify the cancer risk.

ESOPHAGEAL CANCER

Risk assessment

The incidence of esophageal cancer varies widely between countries with a frequency of 2–3 in 100,000 in the UK and North America. In parts of South Africa, the incidence is as high as 50–100 in 100,000. The majority of cases are squamous cell carcinomas. Risk factors for esophageal malignancy include smoking, alcohol, ingestion of spicy foods and reflux. The presence of Barrett esophagus is a further risk factor, with 10% of cases developing adenocarcinoma of the esophagus. Genetic factors do not appear to be important in the etiology of this disease, although familial clustering of Barrett esophagus has been reported. There is a high incidence of esophageal malignancy in parts of China, and research in this population suggest a major mutation underlying the susceptibility to this disease. This study suggests an autosomal recessive inheritance with a sex-specific penetrance. Esophageal malignancy does occur in the inherited condition, tylosis.

Tylosis

(also known as: keratosis palmaris et plantaris with esophageal cancer)

MIM	148500
Clinical features	Late-onset tylosis is associated with esophageal cancer. Tylosis is a non-epidermolytic palmoplantar keratoderma that results in hyperkeratosis of the palms and soles. The risk of esophageal cancer reaches 95% by 65 years of age. Oral leukoplakia is also a feature of the syndrome. The early onset form of tylosis, which starts in infancy, is not associated with esophageal malignancy.
Screening	In known carriers, endoscopy should be performed on a regular basis with biopsies looking for dysplasia. If there is any evidence of dysplasia a prophylactic esophagectomy, replacing the esophagus with part of the colon, is advised. It has been suggested that all affected individuals should undergo prophylactic esophagectomy.
Age of onset	Hyperkeratosis occurs in late childhood. The onset of esophageal cancer is from the age of 10 years.
Inheritance	Autosomal dominant
Chromosomal location	17q24
Gene	*TOC* (tylosis esophageal cancer): loss of heterozygosity at this locus in sporadic tumors suggests that *TOC* is involved in the pathogenesis of sporadic tumors.
Function of protein	Not known
Genetic testing	Work is ongoing to clone the gene. Theoretically, in a large family, testing may be available by linkage.

GASTRIC CANCER

Risk assessment

Risk factors for gastric malignancy include smoking, excess alcohol intake, *Helicobacter pylori* infection and a family history of the disease. Men have a higher risk of gastric malignancy than women (lifetime risk of 1 in 50 vs. 1 in 100).

About 10% of gastric malignancies show familial clustering. Epidemiological studies have demonstrated a two- to three-fold increase in the risk of gastric cancer in first-degree relatives. Both diffuse gastric carcinoma and intestinal gastric cancer have been described in familial forms. Diagnostic criteria have been drawn up for both types of malignancy.

Familial diffuse gastric cancer

Two or more cases of diffuse gastric cancer in first- or second-degree relatives, one of whom is diagnosed under the age of 50 years; or three or more cases of diffuse gastric cancer in first- or second-degree relatives regardless of age of onset.

Familial intestinal gastric cancer

At least two first- or second-degree relatives with intestinal gastric cancer, one of whom is diagnosed before 50 years; or three or more relatives with intestinal gastric cancer at any age.

As well as site-specific gastric malignancies, gastric cancer has been described in familial adenomatous polyposis, hereditary non-polyposis colorectal cancer, Peutz–Jeghers syndrome, Cowden disease and Li–Fraumeni syndrome.

Screening

The diagnosis of gastric malignancy at an early stage greatly increases the chances of survival. Therefore, early detection by a screening program is important in family members at risk of developing gastric cancer. Gastroscopy with multiple biopsies has been suggested. No trials have been undertaken to find the optimum surveillance programs for families with an increased lifetime risk of stomach cancer. However, current practice is to suggest screening intervals of 3 years from 25 years of age if the gastroscopy and biopsies are normal. All at-risk family members should have *Helicobacter* testing and eradication treatment if necessary.

Age of onset   Familial gastric malignancies occur at a younger age than sporadic tumors, generally before the age of 50 years.

Gene   *CDH1*

APC (see FAP)

MLH1 and *MSH2* (see HNPCC)

p53 (see Li–Fraumeni syndrome)

PTEN (see Cowden disease)

STK11 (see Peutz–Jeghers syndrome)

E-cadherin

(also known as: cadherin 1; CDH1)

MIM	192090

Clinical features

Mutations in E-cadherin confer a 70% lifetime risk of gastric malignancy. This was first reported in a large Maori family and this penetrance may be revised as more affected families are identified. The malignancy is generally of the diffuse type and may result in linea plastica.

Screening

Mutation carriers should undergo endoscopy and biopsies on a regular basis. Current research is aimed at increasing the sensitivity and specificity of screening using methods such as chemoendoscopy (the use of methylene blue or indigo carmine for staining) and endoscopic ultrasound.

The use of prophylactic gastrectomy with pathological identification of the duodenal and esophageal mucosa in the surgical specimen has been suggested. While this is effective in preventing gastric cancer, it has a mortality rate of 1–2% and a morbidity rate of acute problems of 10–20%. All patients have long-term problems with weight loss, rapid intestinal transition and diarrhea.

Age of onset

Average age of onset is 38 years, but cases have been described in the third decade.

Inheritance

Autosomal dominant

Chromosomal location

16q22.1

Gene

CDH1: 16 exons spanning 100 kb of genomic DNA encoding a 728 amino acid protein. The majority of mutations are truncating, and they are found throughout the gene. Abnormalities of CDH1 promoter methylation have been described in inherited gastric carcinoma.

Function of protein   Cadherins are transmembrane proteins involved in calcium-dependent cell adhesion. E-cadherin (also known as uvomorulin) is known to be crucial in the formation and maintenance of cell-cell adhesions. It is suggested that mutant E-cadherin loses its Ca^{2+} binding ability which then disrupts cell-cell adhesion. This implies a role for E-cadherin in the suppression of tumor invasion. It has also been suggested that E-cadherin has a tumor suppressor effect.

Genetic testing   Mutation testing is currently on a research basis only

HEAD AND NECK CANCER

Hereditary paraganglioma
(also known as: familial glomus tumor; carotid body tumor)

MIM	168000
Clinical features	Paragangliomas are rare tumors that arise from paraganglionic tissue in the head and neck, which is derived from the neural crest. The parasympathetic paraganglia lie along the course of the parasympathetic nerves, in particular at their intersection with the large vessels. Their function is largely chemoreceptive. The majority of paragangliomas are in the head and neck, with the most common sites being the carotid body, glomus jugulare and glomus vagale. Approximately 4% of tumors are thought to metastasize. The proportion of tumors that are familial is poorly defined: figures range from 5–50%, presumably due to the varying penetrance of the disease. Familial cases are more frequently bilateral and/or multiple and are diagnosed at an earlier age. Whilst the majority of the tumors are benign, pressure effects cause major morbidity due to cranial nerve paralysis and compression of the brain stem.
Screening	Both patients and unaffected at-risk relatives need to be screened by MRI of the head and neck. The majority of paragangliomas have a low growth rate, suggesting that screening every 3 years would be adequate.
Age of onset	15–45 years
Epidemiology	Rare
Inheritance	Autosomal dominant. Consistent with genomic imprinting, transmission of the gene is usually via the paternal line.

Chromosomal location	11q23; *SDHD*
	1q21; *SDHC*
Gene	*SDHD*: succinate dehydrogenase complex, subunit D. A mitochondrial complex II gene comprised of four exons spanning 19 kb of genomic DNA that encode a large subunit of cytochrome b in succinate-ubiquinone oxidoreductase. There is no evidence of imprinting at a molecular level, as demonstrated by biallelic expression in a number of tissues. However, there may be monoallelic expression confined to the paraganglioma cells.
	SDHC: succinate dehydrogenase complex, subunit C. A mitochondrial complex II gene comprised of six exons spanning 35 kb genomic DNA encoding for a small subunit of cytochrome b. There may be another locus, PGL2.
Function of protein	Succinate-ubiquinone oxidoreductase, complex II, is an enzyme in the tricarboxylic acid cycle and the aerobic respiratory chains of mitochondria. It has been suggested that the subunits coded for by *SDHD* and *SDHC* have a role to play in the normal physiology of the carotid body. The loss of either of these subunits due to mutations in *SDHD* or *SDHC* may result in chronic hypoxic stimulation and resultant cellular proliferation. *SDHD* and *SDHC* appear to function as tumor suppressor genes.
Genetic testing	Mutation screening is available on a research basis only

Squamous cell carcinoma of the head and neck

Risk assessment   Tobacco smoking and alcohol consumption are the major risk factors for squamous cell carcinoma of the head and neck (SCCHN). However, evidence is accumulating of a genetic predisposition. A number of studies have now demonstrated familial clustering that is independent of both smoking and alcohol intake.

Second primary tumors are seen in 10–30% of familial clustering cases. First- and second-degree relatives have a four-fold increased risk of developing SCCHN, which can be further increased if the relative smokes and drinks.

Screening   Whilst screening would be difficult, due to the variety of sites affected, it would be worth discussing alteration of lifestyle with first-degree relatives and advising cessation of smoking and drinking.

Gene   It has been suggested that nullizygosity for *GSTT1*—a gene coding for an enzyme involved in detoxification of environmental carcinogens—may predispose to SCCHN in non-smokers exposed to environmental tobacco smoke.

LEUKEMIA

Risk assessment

Leukemias have a bimodal distribution with peaks in children and the elderly. In patients under the age of 15 years, 80% of leukemias are acute lymphocytic leukemias, 17% are acute myeloblastic leukemias and its variants, with chronic granulocytic leukemia making up the remainder (3%). In patients over the age of 15 years, 85% of cases are acute myeloblastic leukemias. Ionizing radiation, chemicals and therapeutic treatments for other malignancies are known to be involved in the etiology of leukemias.

Familial clustering of leukemias has been described, but families invariably share the same environment making the genetic contribution hard to assess. However, there are a number of genetic conditions predisposing to leukemias.

Gene

ATM (see ataxia telangiectasia)

FAA/FAC (see Fanconi anemia)

BLM (see Bloom syndrome)

RBS19 (see Diamond–Blackfan syndrome)

Shwachman syndrome (gene unknown)

WASP (see Wiskott–Aldrich syndrome)

Bloom syndrome

(also known as: BLM; BLS)

MIM	210900
Clinical features	Characterized by pre- and postnatal growth deficiency, hypo- and hyperpigmented skin, increased sun sensitivity, telangiectasia and chromosomal fragility. There is a predisposition to malignancies including esophageal, bowel and hematological tumors. Wilms' tumor has also been described. The oldest recorded survivor was 48 years old but the average age of death due to malignancy is 27 years. The leukemia in BLM presents with leukopenia as opposed to leukocytosis. Male patients are sterile whilst female patients have reduced fertility. BLM appears to have a higher frequency in the Ashkenazi Jewish population.
Screening	Early diagnosis of leukemia does not appear to have an effect on the prognosis of the disease. Given the wide range of solid tumors described with the condition, screening is not thought to be of benefit. A high degree of clinical suspicion is needed when managing patients with BLM.
Age of onset	From 4 years of age
Inheritance	Autosomal recessive
Chromosomal location	15q26.1
Gene	*BLM*: encodes a 1417 amino acid protein.
Function of protein	BLM cells are characterized by a high frequency of sister chromatid exchange, usually a marker for DNA damage. The BLM protein is a member of a family of helicases, the RecQ family. It is located in the nucleus and may be part of a DNA surveillance mechanism.

Mutations result in increased sister chromatid exchange and resultant chromosomal instability.

Genetic testing   Mutation testing is available on a research basis only

Diamond–Blackfan syndrome

(also known as: erythrogenesis imperfecta)

MIM	205900 (105650: autosomal dominant form)
Clinical features	Characterized by hypoplastic macrocytic anemia presenting in the first year of life. Craniofacial anomalies, short stature and abnormal thumbs may be present. There is an increased risk of leukemia. The majority of cases are sporadic, although familial cases are well described.
Screening and management	Patients respond well to steroids. The degree of response is related to the age at presentation, family history and platelet count at presentation. Patients with a family history have a better outcome. Bone marrow transplantation is being used successfully.
Age of onset	70% present by 3 months
Epidemiology	5 in 1,000,000
Inheritance	Autosomal recessive (Autosomal dominant)
Chromosomal location	19q13.2 Also mapped to chromosome 8
Gene	*RPS19* (ribosomal protein S19): consists of 6 exons spanning 11 kb encoding a 145 amino acid protein. Mutations have been found in familial cases, but all patients with the mutation were heterozygotes.
Function of protein	A ribosomal protein
Genetic testing	Mutation testing is available on a research basis only

Shwachman syndrome

MIM 260400

Clinical features Characterized by pancreatic insufficiency (the majority of patients have steatorrhea and low trypsinogen activity) and pancytopenia. Growth retardation is present at birth and skeletal anomalies are common. Short stature is usual. There is a predisposition to myelodysplasia and leukemia.

Screening and management Bone marrow should be monitored for abnormalities; however, not all patients progress from myelodysplasia to leukemia. Bone marrow transplants have been used with variable success.

Age of onset Birth

Inheritance Autosomal recessive

Chromosomal location Linked to chromosome 7, although there is still a large candidate region.

Gene Not known

Function of protein Not known

Wiskott–Aldrich syndrome

(also known as: eczema-thrombocytopenia-immunodeficiency syndrome)

MIM	301000

Clinical features Characterized by a triad of disorders: eczema, thrombocytopenia and immunodeficiency. Children are prone to infections and bloody diarrhea. Abnormalities of the immune system vary between both families and members of the same family, although 60% of patients have a low CD8+ count. Most patients present with clinical signs of thrombocytopenia. Eczema is present in 80% of patients. In the early 1990s the average age of death was 8 years, although the prognosis has now improved with bone marrow transplants. Malignancies of the lymphoreticular system occur in about 13% of patients.

Screening and management Prenatal diagnosis is possible by sexing of the fetus and looking for small platelets if an X-linked gene mutation has been identified within the family. At risk boys should have an FBC undertaken from birth to check for thrombocytopenia. Early bone marrow transplant is now improving the prognosis.

Age of onset The average age at diagnosis is 21 months

Epidemiology 4 in 1,000,000 live male births in the US

Inheritance X-linked recessive

Chromosomal location Xp11.23–p11.22

Gene *WASP* (Wiskott–Aldrich syndrome protein): 11 exons spanning 9 kb of genomic DNA encoding a 501 amino acid protein. Mutations have been found throughout the gene and in X-linked thrombocytopenia.

Function of protein	The WAS protein has a GTPase binding site and interacts specifically with CDC42, which is involved in the regulation of the actin cytoskeleton. T lymphocytes of affected males possess abnormal actin cytoskeletons. It has also been suggested that a mutation results in abnormal coupling of surface immunoglobulins to B cells, resulting in aberrant activation.
Genetic testing	Mutation testing is available on a research basis only

LUNG CANCER

Risk assessment

Lung cancer is the most common malignancy in the developed world and its incidence is gradually increasing. The etiology of lung cancer is strongly associated with cigarette smoking, although other environmental factors have been implicated including occupational respiratory substances such as asbestos and general pollutants in the air. Adenocarcinoma and alveolar cell carcinoma account for only a small proportion of lung cancers and are known to be unassociated with smoking. There have been reports of familial clustering of the disease. A recent study suggests that family history of the disease is associated with early onset. Adenocarcinoma of the lung is known to be associated with Li–Fraumeni syndrome. Research studies are ongoing to search for loci involved in early-onset lung cancer.

LYMPHOPROLIFERATIVE DISORDERS

Risk assessment

Non-Hodgkin lymphoma is a heterogeneous group of diseases that may have a familial component. There is a definite association between genetic conditions causing immunodeficiency and the development of lymphomas. Non-Hodgkin lymphoma is more common in these conditions than Hodgkin lymphoma. There is a four-fold increased risk of leukemias and lymphomas among first-degree relatives of patients with a non-Hodgkin lymphoma. Familial cases of Hodgkin lymphoma have been described although it is difficult to determine whether these are related to genetic or environmental factors. Siblings of young patients are known to have a seven-fold excess risk of the disease. There is no excess risk for siblings of older patients with Hodgkin lymphoma. Concordance studies suggest that genetic susceptibility underlies the development of Hodgkin disease in young people.

Gene

ATM (see ataxia telangiectasia)

WASP (see Wiskott–Aldrich syndrome)

CHS1 (LYST) (see Chédiak–Higashi syndrome)

MYO5A (see Griscelli syndrome)

SH2D1A (see X-linked lymphoproliferative disorder)

DSCR (see Down syndrome critical region 22q23)

Chédiak–Higashi syndrome

(also known as: CHS)

MIM	214500

Clinical features
Characterized by decreased pigmentation of hair and eyes, eosinophilic inclusion bodies in the myeloblasts and promyelocytes of the bone marrow, mild bleeding tendency, immunological deficiency and progressive neurological degeneration. About 85–90% of patients develop a lymphoproliferative disorder characterized by lymphohistiocytic infiltrates, hepatosplenomegaly, lymphadenopathy, pancytopenia and jaundice. Death usually occurs before the age of 7 years. Those that do not develop the lymphoproliferative phase develop neurological manifestations.

Screening and management
Clinical management depends upon treatment of infections and hemostatic measures. Histological diagnosis is based on the presence of giant granular inclusion bodies, giant lysosomes and giant melanosomes. Bone marrow transplants have been used for treatment of the condition.

Age of onset
Hypopigmentation from birth

Inheritance
Autosomal recessive

Chromosomal location
1q42.1–q42.2

Gene
CHS1: encodes a 3801 amino acid protein. Homozygosity for null mutant alleles results in severe forms of the disease. Homozygosity for missense mutant alleles causes a milder phenotype.

Function of protein
The protein is likely to be involved in lysosomal intracellular trafficking. Cytotoxic T lymphocyte associated antigen 4 (CTLA4) is involved in the regulation of T cell activation. It has been suggested

that aberrant presentation of this antigen, due to abnormal lysosomal trafficking, may result in lymphoproliferative disorders.

Genetic testing   Mutation testing is available on a research basis only

Griscelli syndrome

MIM   214450

Clinical features   Similar in phenotype to Chédiak–Higashi syndrome (CHS), with
                            variable hypopigmentation of the skin and hair, and cellular immune
                            deficiency. Children go on to develop the same accelerated
                            lymphoproliferative disorder as seen in CHS but there are no
                            neurological complications. The diagnostic differences are in the
                            histology of hair, finding large clumps of pigment in the hair shaft
                            and accumulation of mature melanosomes in the melanocytes.

Screening and   Management is the same as for CHS. Bone marrow transplantation
management   is the only treatment for the lymphoproliferative disorder.

Age of onset   Hypopigmentation at birth

Inheritance   Autosomal recessive

Chromosomal location   15q21

Gene   MYO5A: encodes myosin Va.

                            RAB27A: five coding exons spanning 65kb. It encodes a 221 amino
                            acid rab protein.

Function of protein   The class V myosins are essential for the T cell response and are
                            involved in the movement of organelles. Mutations in MYO5A result
                            in Griscelli syndrome which has characteristic neurological, but no
                            immune, features. Rab proteins are involved in the GTP-binding
                            pathways. RAB27A is probably involved in the immune system, as
                            patients with mutations within the gene develop hemophagocytic
                            syndrome.

Genetic testing   Mutation testing is available on a research basis only

X-linked lymphoproliferative syndrome

(also known as: Duncan disease)

MIM	308240
Clinical features	Characterized by progressive combined variable immunodeficiency with benign or malignant proliferation of lymphocytes and histiocytes, and abnormalities of immunoglobulins. Exposure to Epstein–Barr virus (EBV) results in fatal infectious mononucleosis (50%), B-cell lymphoma (25%) or agammaglobulinemia (25%). There is 100% mortality by the age of 40 years. The lymphomas are most often Burkitt lymphomas affecting the ileocecal region.
Screening and management	Bone marrow transplantation appears to restore normal host immune response to EBV. Transplantation before the age of 15 years seems to confer a better prognosis.
Age of onset	Average age of onset is 2.5 years
Epidemiology	Rare, 240 cases identified worldwide between 1981 and 1991
Inheritance	X-linked
Chromosomal location	Xq25
Gene	*SH2D1A* (also known as *SAP*: SLAM- Associated Protein) four exons encode a 128 amino acid protein.
Function of protein	It is thought to control the signal-transduction pathways involved in the stimulation of T and B cells. Evidence suggests that this gene product is important in maintaining a normal host response to EBV infection.
Genetic testing	Not available

OVARIAN CANCER

Risk assessment

Ovarian cancer is the sixth most common female cancer in the UK. The lifetime risk of epithelial ovarian cancer is about 1 in 70. The major risk factor for ovarian cancer is a family history of the disease. About 10% of ovarian cancer is due to an inherited predisposition. Having an affected first-degree relative increases a woman's chance of developing the disease by three-fold. If a woman has two affected close relatives, the risk is increased to about 1 in 10. If there are more cases of ovarian malignancy, especially with an early age of diagnosis, then the risks are those of an autosomal dominant family. Any breast cancers within the family need to be taken into account when assessing risk. The clearly autosomal dominant predisposition to ovarian cancer is seen in epithelial cancers. Mutations in *BRCA1/2* account for the majority of families with inherited ovarian cancers, even those families without cases of breast cancer. However, there may be a further breast/ovarian cancer predisposition gene that has not yet been mapped or cloned. Non-epithelial ovarian tumors are seen in Peutz–Jeghers and Gorlin syndromes.

Screening

Ovarian cancer is curable if diagnosed at an early stage. Therefore, the aim of any screening program is to detect ovarian malignancies at Stage 1. The current suggestion is that women whose lifetime risk of developing ovarian cancer is ≥1 in 10 should be offered screening. This is usually with CA125 and transvaginal ultrasound examination with color Doppler flow on an annual basis from around 30 years of age. The efficacy of this screening protocol is still being assessed. CA125 screening alone only detects Stage 1 disease in 50% of cases. An observational trial is underway in the UK to assess the efficacy of transvaginal ultrasound in conjunction with CA125 measurements in a high-risk population. Families with two or more cases of ovarian cancer are at increased risk of developing breast cancer, therefore, both affected and at-risk members should be offered mammograms on either an annual or 18-monthly basis according to age (see breast cancer).

Prophylactic oophorectomy has been suggested to prevent ovarian cancer although there may still be a risk of peritoneal cancer. Primary peritoneal cancer does occur with an increased incidence in *BRCA1* families and therefore CA125 levels should still be measured even after prophylactic oophorectomy in high-risk families.

It has been shown in known *BRCA1* mutation carriers that the combined oral contraceptive decreases the risk of ovarian cancer by up to 60%. However, it must be balanced with the increased risk of breast cancer.

Age of onset	From as early as the fourth decade in some families
Gene	*BRCA1/2* (see breast cancer)
	MLH1 and *MSH2* (see HNPCC)
	STK11 (see Peutz–Jeghers syndrome)

PANCREATIC CANCER

Risk assessment

The incidence of pancreatic cancer is increasing. About 6% of patients with pancreatic cancer are thought to have a first-degree relative diagnosed with the disease. Both autosomal dominant and autosomal recessive families have been described—although the autosomal recessive cases may be explained by autosomal dominant inheritance with reduced penetrance.

A collaborative study is underway to localize genes involved in hereditary pancreatic cancer.

Pancreatic cancer may be part of other inherited cancer syndromes and is seen in Li–Fraumeni syndrome, with *BRCA2* mutations and in familial atypical multiple mole-melanoma (FAMMM) syndrome. Hereditary pancreatitis also predisposes to pancreatic cancer.

Screening and management

Pancreatic cancer is difficult to screen for. It has been suggested that ultrasound may be useful in the screening of high-risk patients, i.e. those families with clear inheritance of the disorder. Endoscopic retrograde cholangiopancreatography (ERCP) is currently the diagnostic tool of choice. However, due to the invasive nature of the procedure, it is unlikely to be useful in long-term screening. The more recent procedure of magnetic resonance cholangiopancreatography is non-invasive. This appears to have a similar sensitivity and specificity to ERCP in a symptomatic diagnostic setting and may be worth evaluation as a screening tool for high-risk families.

Gene

TP53 (see Li–Fraumeni syndrome)

CDKN2 (see FAMMM syndrome)

BRCA2 (see breast cancer)

PRSS1 (see hereditary pancreatitis)

Pancreatic cancer

Hereditary pancreatitis

(also known as: HPC)

MIM	167800
Clinical features	Characterized by acute episodes of abdominal pain, massively elevated serum amylase levels and fever. There is an increased incidence of pancreatic cancer, with a cumulative lifetime risk of 40%. Some patients have calcium lithiasis.
Screening and management	Clinical management is symptomatic. Screening for malignancy should be considered.
Age of onset	The average age for the first episode of pancreatitis is 14 years
Epidemiology	Over 50 families described worldwide
Inheritance	Autosomal dominant
Chromosomal location	7q35
Gene	*PRSS1*: 5 exons, encoding for trypsin 1. About 50% of families screened have mutations.
Function of protein	Trypsin activation usually occurs in the stomach. If it is activated in the pancreas it is inhibited by a (normally finite supply of) trypsin inhibitor (pancreatic secretory trypsin inhibitor—PSTI). However, when the amount of trypsin exceeds PSTI levels a negative feedback mechanism begins. The proenzymes mesotrypsin and enzyme Y are activated and, in the normal situation, inactivate the trypsin molecules. However, in HPC, the trypsin cleavage site is mutated and mesotrypsin and enzyme Y can not inactivate it. This leads to autodigestion of the pancreas by trypsin, and subsequent pancreatitis.
Genetic testing	Mutation testing is available on a research basis only

PROSTATE CANCER

Risk assessment

Prostate cancer is the most common form of cancer affecting males in the US. It has been estimated that 9% of Caucasian men will develop the disease in their lifetime. Familial clustering of prostate cancer has been observed. It has been suggested that first-degree relatives of patients with prostate cancer have a three-fold higher risk of developing the disease—risk is inversely proportional to the relative's age at diagnosis. The risk also increases with the number of affected cases within a family—a man with three affected relatives has an 11-fold increased risk. Some of the familial clustering of cases could be explained by a rare, but highly penetrant, allele of autosomal dominant inheritance. Genetic susceptibility may account for up to 10% of all cases of prostate cancer.

Screening

Screening for prostate cancer is problematic. Prognosis is improved by early diagnosis, but as yet there are few data with regard to screening and reduction in mortality. It has been suggested that screening may increase morbidity. Prostate-specific antigen (PSA) may be a useful tool in a high-risk population. Chemoprevention agents are also being considered within a high-risk population.

Age of onset

Familial prostatic malignancies occur at a younger age than sporadic tumors.

Inheritance

Autosomal dominant inheritance seems the most plausible model. It has been estimated that a dominant allele would occur with low frequency and account for about 9% of prostate cancers in patients under the age of 85 years, and up to 43% of prostate cancer diagnosed at 55 years or younger. It has also been suggested that some studies are consistent with autosomal recessive or X-linked inheritance.

Gene

BRCA1 and *BRCA2*: both confer an increased lifetime risk of prostate cancer. No familial prostate cancer genes have been cloned to date although a number of loci have been identified using linkage analysis.

HPC1 (1q24–25): most common in families with a mean age of diagnosis of less than 65 years, with four or more relatives with the disease, and those with more advanced-stage disease May account for 6% of families with hereditary prostate cancer.

PCAP (1q42.2–43): a single report of linkage with other studies failing to corroborate this locus.

HPC20 (20q13): a report of linkage to this region in families with less than five members affected, later age at onset and no male-to-male transmission. The mode of inheritance could not be determined.

HPCX (Xq27–28): best evidence in a subset of families with no male-to-male transmission.

Genome scans in families with prostate cancer are revealing other regions of interest, and further studies are being undertaken.

RENAL CELL CARCINOMA

Risk assessment

About 2% of renal cell carcinomas are familial. As with other inherited cancers, age of onset is earlier and the tumor is more likely to be multifocal or bilateral. There are two different histological types of renal cell carcinoma; clear cell (non-papillary type) and papillary type, both of which have different inherited forms.

The majority of familial cases of clear cell carcinoma are due to von Hippel–Lindau (VHL) disease. However, there is a familial form of clear cell carcinoma that does not have any VHL disease features. Fifty-six percent of patients within these families develop a malignancy before the age of 50 years. *VHL* mutations have been excluded within these families and the gene involved has not been mapped.

Papillary renal cell carcinoma has been described in a familial form and is due to mutations in the *MET* oncogene. This type of cancer is also seen in tuberous sclerosis and Birt–Hogg–Dubé syndrome.

Screening

Screening for renal cancers usually involves ultrasound screening of the kidneys on an annual basis from 15 years in all familial forms of the disease. This is because hereditary renal cancers are associated with an earlier age of onset than renal cancers in the general population. A young person with renal clear cell carcinoma should have an extensive family history taken and an ophthalmic examination to rule out retinal angiomas. DNA should be screened for mutations in the *VHL* gene. If there are any neurological signs, an MRI of the posterior fossa should be undertaken. If all these investigations are normal, it is likely that the family have a familial renal clear cell carcinoma rather than VHL disease.

Age of onset

Earlier age of onset than in the general population (6th and 7th decade).

Gene   *VHL* (see von Hippel–Lindau disease)

TSC2 (see tuberous sclerosis)

MET (see familial papillary renal cell carcinoma)

Birt–Hogg–Dubé syndrome

(also known as: fibrofolliculomas with trichodiscomas and acrochordons)

MIM	135150
Clinical features	Characterized by fibrofolliculomas, trichodiscomas and acrochordons (small papules distributed over the face, neck and upper trunk). Lipomas, collagenomas and oral fibromas have also been described. Renal tumors include renal oncocytomas (benign tumors) and papillary renal cell carcinomas. Mutations in both *VHL* and *MET* genes have been excluded.
Screening	Renal ultrasound has been suggested on an annual basis from 40 years of age.
Age of onset	Renal tumors occur at a similar age as that in the general population (6th–7th decade).
Inheritance	Autosomal dominant
Chromosomal location	Mapped to 17p11.2
Gene	Unknown
Function of protein	Unknown

Familial papillary renal cell carcinoma

MIM   164860

Clinical features   This condition is characterized by multifocal bilateral papillary cell renal cancer.

Screening   Ultrasound of the kidneys in affected and at-risk relatives on an annual basis.

Age of onset   Generally younger than in the general population (6th–7th decade)

Inheritance   Autosomal dominant

Chromosomal location   7q31

Gene   *MET*: mutations not found in sporadic papillary renal cell carcinoma.

Function of protein   *MET* is a member of the tyrosine kinase family of oncogenes. Mutations in the *MET* proto-oncogene result in activation of MET protein, which leads to altered cell growth.

Genetic testing   Mutation testing is available on a research basis only

Wilms tumor

(also known as: nephroblastoma)

MIM	194070
Clinical features	Wilms tumor is one of the most common forms of childhood malignancy. A proportion of tumors are familial, with about 1% having a family history of the disease. A higher proportion of familial cases have bilateral disease. Wilms tumors are associated with Beckwith–Wiedemann syndrome, hemihypertrophy and aniridia as part of the WAGR syndrome consisting of Wilms tumor, Aniridia, Genitourinary abnormalities and mental Retardation. The WAGR syndrome represents a contiguous gene defect. In children with aniridia, the Wilms tumors are more likely to be bilateral and occur at a younger age than in children with Wilms tumor without aniridia.
Screening	In families with inherited Wilms tumor, or in children with aniridia, abdominal examination and ultrasound should be performed in the neonatal period, then at 3-monthly intervals for the first 3 years, then biannually until the age of 8 years. Annual ultrasound should then be considered until 12 years of age.
Age of onset	80% of patients present by the age of 5 years
Epidemiology	1 in 10,000 tumors are sporadic. About 1% of all cases are familial.
Inheritance	Autosomal dominant with reduced penetrance
Chromosomal location	11p13 (*WT1*)
	11p15.5 (*WT2*)
	16q (*WT3*)
	17q12–q21 (*WT4, FWT1*)

Gene	*WT1*: 10 exons encoding four different transcripts through alternative splicing. Mutations in *WT1* have also been found in Denys–Drash syndrome and isolated diffuse mesangial sclerosis.
	WT2: not cloned
	WT3: not cloned
	WT4: not cloned
Function of protein	The WT1 protein is a transcription factor expressed within the developing kidney, fetal gonad, genital ridge and mesothelium. It is involved in the development of the genitourinary tract and is important in cell differentiation, in particular differentiation of the podocyte.
Genetic testing	Mutation testing is available on a research basis only

SKIN CANCER

Risk assessment

There are many different types of genetic predisposition to skin tumors. The scope of this book is to concentrate on malignant conditions. For other conditions please consult 'Genetics for dermatologists' (from the same ReMEDICA series, ISBN 1 901346 10 2). The major types of malignant skin lesion are melanoma, squamous cell carcinoma and basal cell carcinoma. A number of inherited conditions with a predisposition to neoplasia have cutaneous manifestations and have been described elsewhere in the text.

Basal cell carcinoma

(also known as: BCC)

Clinical features
Sun-exposed skin is most affected by this common tumor, which rarely occurs under the age of 40 years. As with other skin tumors, the incidence of BCC has increased over the last few years. Inherited conditions such as Gorlin syndrome, xeroderma pigmentosum, albinism, Bazex syndrome and Rombo syndrome predispose to BCCs. Screening for BCC involves regular skin examination and removal of suspicious lesions. As with any predisposition to skin cancer, patients should be advised about protection from the sun.

Gene
PTCH (see Gorlin syndrome)

XP (see xeroderma pigmentosum)

TYR (see oculocutaneous albinism type I)

OCA2 (see oculocutaneous albinism type II)

Bazex syndrome

(also known as: follicular atrophoderma and basal cell carcinoma; BZX)

MIM	301845
Clinical features	Characterized by follicular atrophoderma that look like 'ice pick' marks to the hands and face. Hair is sparse and has a twisted and flattened appearance on scanning electron microscopy. Basal cell carcinomas occur frequently. BZX may be allelic with Gorlin syndrome but this has not been proven.
Age of onset	Skin lesions develop from 15 years
Epidemiology	Less than 20 families reported
Inheritance	X-linked recessive
Chromosomal location	Xq24–q27
Gene	Not known
Function of protein	Not known

Epidermolysis bullosa dystrophica

(also known as: RDEB)

MIM	226600
Clinical features	Bullae develop from birth, mainly affecting sites of friction, and also occur on mucosal surfaces. Scarring predisposes to squamous cell carcinoma. Multiple allelic forms probably exist within the classification of RDEB. Severe cases may develop to acquired syndactyly and esophageal stricture.
Age of onset	Birth
Inheritance	Autosomal recessive
Chromosomal location	3p21.3
Gene	*COL7A1*: 118 exons encoding pro-α1 chain 7 collagen. The more severe form generally has truncating mutations, whereas the milder type often has missense mutations.
Function of protein	This protein is involved in the formation of the fibrils anchoring the lamina densa to the papillary dermis. Mutations result in abnormal conformation of the fibrils or a marked decrease in protein levels, leading to abnormal anchoring and resultant bullae.
Genetic testing	Mutation testing is available on a research basis only

Familial atypical mole-malignant melanoma syndrome

(also known as: FAMMM)

MIM	155600

Clinical features The majority of families at high risk of developing melanomas (four or more melanoma cases) are due to mutations in *CDKN2*. There is an increased risk of melanomas in this condition with 26% developing metachronous or synchronous melanoma. The lifetime risk of melanoma may be as high as 100% in some families. The age at diagnosis is about 10 years below that of the general population. The lesions tend to be thinner with a lower level of invasion. Some families show an increased incidence of intraocular melanoma. Mutations in *CDKN2* also confer an increased risk of pancreatic and breast cancer.

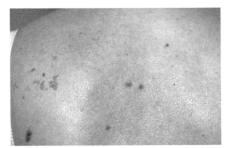

Dysplastic nevi on the back of a patient with malignant melanoma.

Site of removal of melanoma.

Screening	Screening is as previously described, with family members with dysplastic nevi or melanomas being examined on a 6-monthly basis. High level protection sun cream should be used and protective clothing worn. All children should be assumed to be at risk and should therefore be protected against sunburn. Any suspicious lesions should be excised, and any changing dysplastic nevi should be biopsied. As well as general education, family members should be taught self-examination. In particular, they should be warned that hormonal changes, such as pregnancy, may also have an effect on melanomas and dysplastic nevi.
Age of onset	Generally in the third decade, but 10% develop before the age of 20 years.
Inheritance	Autosomal dominant. Penetrance varies between 40–100%.
Chromosomal location	9p21 (*CDKN2*) 1p36 (*CMM1*) gene not yet cloned
Gene	*CDKN2*: 3 coding exons, with the alpha transcript encoding p16. Mutations in this gene account for approximately one-third of high risk melanoma families. Somatic mutations are found in a number of different cell lines.
Function of protein	Wild-type p16 arrests cells in late G1. p16 is known to interact with cyclin D and is involved in the same pathway as the retinoblastoma gene. It has been suggested that p16 acts as a tumor suppressor gene: mutant p16 fails to affect the cell cycle and therefore lacks tumor suppressor properties.
Genetic testing	Mutation testing is available in a few families only—selection is based on family history. Predictive testing may be used within these families to direct surveillance.

Hyperkeratosis lenticularis perstans

(also known as: Flegel disease, HLP)

MIM	144150
Clinical features	Hyperkeratosis occurs mainly over the dorsum of the foot and the leg. There is a high incidence of both basal and squamous cell carcinoma.
Age of onset	Skin lesions develop from the third decade
Epidemiology	Very rare: about four families described in the literature.
Inheritance	Autosomal dominant
Gene	Not known
Function of protein	Not known

Melanoma

Risk assessment

The incidence of melanoma is rising. Risk factors include exposure to sun, pale skin that burns easily, a history of severe sunburn in childhood and the presence of large numbers of nevi or of dysplastic nevi. Individuals with dysplastic nevi, but no family history of melanoma, have a four- to ten-fold increased risk of melanoma. Approximately 12% of melanoma cases have a family history. If an individual with dysplastic nevi has a family history of melanoma, the risk of the disease increases by 100- to 400-fold. In familial cases, as with other inherited predispositions to malignancy, the melanomas occur at a younger age than in the general population. Familial melanomas are clinically and histologically similar to non-familial melanomas. However, the distribution of nevi is atypical in familial cases, occurring on the buttocks, scalp, palms and soles as well as sun-exposed areas.

Screening

Individuals with melanomas or dysplastic nevi should have their skin examined on a regular basis every 6–12 months to detect any changes in dysplastic nevi. First-degree relatives in a melanoma-prone family should have their skin checked on an annual basis from teenage years onwards. All patients should be educated to keep out of the sun as much as possible, to wear high-factor protection sun cream and to wear protective clothing. Self-examination of nevi should be encouraged, looking for change in size, shape, color, whether the nevi are raised and whether there is alteration of sensation over the nevus and surrounding skin.

Age of onset

10% of cases occur before 20 years of age

Gene

CDKN2 (p16)

CMM1: linked to 1p36

CMM3: linked to 6p

p53 (see Li–Fraumeni syndrome)

Oculocutaneous albinism type I

(also known as: albinism I, tyrosinase-negative; OCA1)

MIM	203100
Clinical features	Characterized by the absence of pigment in hair, skin and eyes. As a result, the skin develops actinic changes that may predispose to both BCC and squamous cell carcinoma of the skin. The majority of patients have nystagmus.
Screening	Examination of the skin on a regular basis and protection from the sun. Prenatal diagnosis has been made using an electron microscopic L-DOPA (L-3,4,-dihydroxyphenylamine) reaction test on fetal skin at 20 weeks gestation.
Age of onset	Birth
Epidemiology	1 in 10,000
Inheritance	Autosomal recessive
Chromosomal location	11q14–q21
Gene	*TYR*: 5 exons spanning 50 kb of genomic DNA encoding tyrosinase. Mutations in *TYR* account for 90% of mutations in OCA1 Caucasians. Mutations are clustered in exons 1 and 4.
Function of protein	Tyrosinase aggregates within melanosomes and is involved in the conversion of tyrosine to melanin. Mutations in the *TYR* gene result in loss of melanin production.
Genetic testing	Mutation testing is available on a research basis only

Oculocutaneous albinism type II

(also known as: albinism II; tyrosinase-positive; OCA2)

MIM	203200
Clinical features	Clinically similar to OCA1. Characterized by the absence of pigment in hair, skin and eyes. There may be pigmented nevi present. In the black population, pigmented spots develop on the skin and the hair is yellow. The skin develops actinic changes that may predispose to both BCC and squamous cell carcinoma of the skin. The incidence of skin cancer is lower in families with darkly pigmented patches (ephelides). The majority of patients have nystagmus.
Screening	Examination of the skin on a regular basis and protection from the sun
Age of onset	Birth
Epidemiology	1 in 35,000 in the UK, 1 in 2000 in Africa.
Inheritance	Autosomal recessive
Chromosomal location	15q11.2–q12
Gene	OCA2: 25 exons spanning 250–600 kb. Encodes an 838 amino acid protein melanosome membrane protein.
Function of protein	The protein transports melanin across the melanosome membrane. If it is not functioning, melanin is unable to enter the melanosome.
Genetic testing	Mutation testing is available on a research basis only

Rombo syndrome

MIM   180730

Clinical features   Characterized by cyanotic redness of the lips and hands, follicular
                         atrophy of the skin, and telangiectasia. Whitish papules of the skin
                         develop in adulthood. Basal cell carcinoma is a frequent
                         complication.

Age of onset   Skin changes from 7 years

Epidemiology   Only one family described

Inheritance   Autosomal dominant

Gene   Not known

Function of protein Not known

Rothmund–Thomson syndrome

(also known as: poikiloderma atrophicans and cataract; RTS)

MIM	268400
Clinical features	Characterized by skin lesions including atrophy, pigmentation and telangiectasia. The skin rash is usually sun-sensitive. Other features include juvenile cataracts, hypogonadism, congenital bone defects (75% of cases), saddle nose and abnormal hair growth. Osteosarcoma has been described in 7% of cases. There is an increased risk of skin cancer.
Screening and management	Patients should avoid sun exposure and use sunscreens with UVA and UVB protection. Annual dermatological examination and regular ophthalmology appointments for screening for cataracts should be undertaken. Patients should be made aware of the risk of osteosarcoma.
Age of onset	Skin lesions present within the first year
Epidemiology	Less than 200 families described
Inheritance	Autosomal recessive
Chromosomal location	8q24.3
Gene	*RECQL4*: encodes a 1208 amino acid DNA helicase. Mutations are not found in all cases.
Function of protein	The protein is a DNA helicase involved in unwinding double-stranded DNA during DNA replication. The mutated gene encodes a truncated protein that either does not function, or functions abnormally.
Genetic testing	Mutation testing is available on a research basis only

Squamous cell carcinoma of the skin

Risk assessment

Fair-skinned people are most affected by squamous cell carcinoma, especially in areas exposed to sunlight. This is why albinos are at a high risk of developing the disease. A number of genetic conditions predispose to squamous cell carcinoma, including epidermolysis bullosa, hyperkeratosis lenticularis perstans and Rothmund–Thomson syndrome.

Gene

XP (see xeroderma pigmentosum)

COL7A1 (see epidermolysis bullosa dystrophica)

RECQL4 (see Rothmund–Thomson syndrome)

THYROID CANCER

Risk assessment

Thyroid malignancy is common and occurs more frequently in women than in men. Around 60% of all thyroid cancers are papillary, 15% follicular, 15% anaplastic and the remainder are medullary thyroid cancers. Familial predisposition to papillary and medullary thyroid carcinoma has been described, with the majority of familial medullary cell cancers being associated with MEN2A and MEN2B. A pure familial medullary thyroid carcinoma has been described in association with *RET* mutations.

Papillary cell carcinoma of the thyroid has been described as part of familial adenomatous polyposis and Cowden disease. A pure familial form has been described with younger age at onset and frequent bilateral involvement. These tumors appear to be more aggressive than the sporadic forms of papillary carcinoma and may be due to a rearrangement of the *RET* oncogene. At least one large family has had *RET*, *APC*, *MET* and *PTEN* excluded by microsatellite analysis. Within the general population, between 3–6% of patients will have an affected relative.

Screening

Unaffected relatives who are at risk of developing the disease are given basal and stimulated calcitonin, serum calcium and parathyroid level screening and 24 h urine collection to test for catecholamines on an annual basis from 6–35 years—if genetic screening is not available. Known mutation carriers are offered prophylactic thyroidectomy from 6 years. With a family history of papillary carcinoma of the thyroid in more than one individual, screening by clinical examination of the neck and regular ultrasound scans has been advocated. The frequency at which this screening should be undertaken has not been established.

Age of onset

Familial malignancies are usually seen at an earlier age than in the general population.

Gene   *RET* (see MEN2A and MEN2B)

                      APC (see FAP)

                      PTEN (see Cowden disease)

3. Chromosomal abnormalities in cancer

Introduction

Structural and/or numerical chromosomal abnormalities are common in neoplasia, although the role of these abnormalities in tumorigenesis remains unclear in many cases. The non-random association between certain chromosomal regions and specific tumor types has led to some promising areas of research that may provide proof of a causative relationship. In some tumors, a specific chromosomal abnormality is studied as a prognostic factor.

The range of chromosomal abnormalities described in cancer is vast. In a review published in Nature Genetics (April, 1997), Mitelman *et al.* analyzed all cancer-associated rearrangements that had been reported up until 1996. Of the 26,523 cases reviewed, 215 were balanced and 1588 were unbalanced chromosomal rearrangements—there was a total of 75 tumor types.

Table 1. Frequent chromosomal rearrangements seen in specific conditions.

Rearrangement	Neoplasia
t(1;3)(p36;q21)	AML
t(1;22)(p13;q13)	AML
t(1;19)(q23;p13)	ALL
t(2;13)(q35;q14)	Rhabdomyosarcoma
inv(3)(q21;q26)	AML
t(3;8)(p21;q21)	Salivary gland adenoma
t(4;11)(q21;q23)	ALL
t(8;14)(q24;q32)	ALL, NHL
t(8;21)(q22;q22)	Ewing's sarcoma
t(9;22)(q34;q11)	CML Ph+
inv(10)(q11;q21)	Thyroid adenocarcinoma
t(12;16)(q13;p11)	Liposarcoma
t(12;14)(q15;q22)	Uterine leiomyoma

ALL: acute lymphoblastic leukemia; AML: acute myelogenous leukemia; CML: chronic myelogenous leukemia; NHL: non-Hodgkin's lymphoma.

The balanced rearrangements seem to be more disease specific. Table 1 lists some examples of the more frequent recombinations. (For a comprehensive review, see Mitelman F, Mertens F, Johansson B. A breakpoint map of recurrent chromosomal rearrangements in human neoplasia. Nature Genetics (1997) Apr;15 Spec No:417–74.)

The Philadelphia chromosome

The first abnormality to be described was the Philadelphia chromosome (Ph) in chronic myelogenous leukemia (CML). This is a translocation between chromosomes 9 and 22 at bands q34 and q11—written as t(9;22)(q34;q11). About 95% of patients with CML have this chromosomal rearrangement, which also occurs in 25–30% of adult and 2–10% of childhood acute lymphoblastic leukemia (ALL). In this translocation, the *ABL1* proto-oncogene is moved from its position on chromosome 9 to a region known as the break-point cluster region (bcr) on chromosome 22. *ABL1* encodes a cytoplasmic and nuclear protein receptor kinase (c-Abl) involved in cell differentiation, division, adhesion and stress response. Activity of the c-Abl protein is negatively regulated by its SH3 domain. The translocation to chromosome 22 results in the deletion of the SH3 domain, converting *ABL1* into an oncogene. The chimeric gene then has functions not seen in the wild-type. Most patients with chronic phase CML have the Philadelphia chromosome as their only chromosomal abnormality. However, as the disease progresses other chromosomal abnormalities are noted. These may precede clinical or hematological changes of progression and, therefore, cytogenetic analysis is used to determine prognosis.

Other chromosomal abnormalities and their uses

The majority of studies on chromosomal abnormalities that result in oncogenic conversion of genes have been undertaken in hematology. As well as the CML model described above, other translocations have been suggested. Many of the translocations in lymphocytic leukemias involve the immunoglobulin or T cell receptor genes. If a cellular gene is juxtaposed to either the immunoglobulin or T cell receptor loci, there appears to be altered transcription of the cellular gene. This then results in altered expression of the cellular gene, resulting in abnormal cell differentiation or proliferation.

Chromosomal aberrations are being used as part of the diagnostic classification of a number of groups or subgroups of hematological disorders. These subgroups are characterized by particular chromosomal changes, and are the basis of the Morphologic, Immunologic and Cytogenetic (MIC) classification. For example, acute myelogenous leukemia has 10 subgroups characterized by specific chromosomal abnormalities—subgroup 1, M$_2$ acute myeloblastic leukemia, has a translocation t(8;21)(q22;q22) that appears almost exclusively in this subgroup.

The characterization of cytogenetic anomalies in solid tumors is not as advanced as in the hematological malignancies. In some tumors, such as retinoblastoma and Wilms tumor, the discovery of chromosome deletions has aided in the localization and cloning of the gene responsible for the condition. Some rearrangements are common, for example 92% of cases of Ewing's sarcoma have a translocation t(11;22)(q24;12).

Whilst it is obvious that some chromosomal changes, such as those in CML and those involving tumor suppressor genes, are involved in the pathogenesis of malignancy, the importance of many other chromosomal rearrangements in tumorigenesis is unknown. For example, lipomas have a characteristic chromosomal abnormality but do not demonstrate a malignant phenotype. There is also uncertainty about the role of secondary chromosomal changes and their effect on the ability of a tumor to invade and metastasize. There is still a large amount of work to be undertaken in the field of cancer cytogenetics.

4. Glossary

A

Adenine (A)   One of the bases making up **DNA** and **RNA** (pairs with **thymine** in DNA and **uracil** in RNA).

Agarose gel   See **electrophoresis**
electrophoresis

Allele   One of two or more alternative forms of a **gene** at a given location (**locus**). A single allele for each locus is inherited separately from each parent. In normal human beings there are two alleles for each locus (**diploidy**). If the two alleles are identical, the individual is said to be **homozygous** for that allele; if different, the individual is **heterozygous**.

For example, the normal **DNA** sequence at **codon** 6 in the beta-globin gene is GAG (coding for glutamic acid), whereas in sickle cell disease the sequence is GTG (coding for valine). An individual is said to be heterozygous for the glutamic acid → valine **mutation** if he/she possesses one normal (GAG) and one mutated (GTG) allele. Such individuals are **carriers** of the sickle cell gene and do not manifest classical sickle cell disease (which is **autosomal recessive**).

Allelic heterogeneity   Similar/identical **phenotypes** caused by different **mutations** within a **gene**. For example, many different mutations in the same gene are now known to be associated with Marfan's syndrome (*FBN1* gene at 15q21.1).

Amniocentesis   Withdrawal of amniotic fluid, usually carried out during the second trimester, for the purpose of prenatal diagnosis.

Amplification   The production of increased numbers of a **DNA** sequence.

1. *In vitro*
In the early days of recombinant DNA techniques, the only way to amplify a sequence of interest (so that large amounts were available

for detailed study) was to **clone** the fragment in a vector (**plasmid** or phage) and transform bacteria with the recombinant vector. The transformation technique generally results in the 'acceptance' of a single vector molecule by each bacterial cell. The vector is able to exist autonomously within the bacterial cell, sometimes at very high copy numbers (e.g. 500 vector copies per cell). Growth of the bacteria containing the vector, coupled with a method to recover the vector sequence from the bacterial culture, allows for almost unlimited production of a sequence of interest. Cloning and bacterial propagation are still used for applications requiring either large quantities of material or else exceptionally pure material.

However, the advent of the **polymerase chain reaction** (PCR) has meant that amplification of desired DNA sequences can now be performed more rapidly than was the case with cloning (a few hours cf. days), and it is now routine to amplify DNA sequences 10 million fold.

2. *In vivo*
Amplification may also refer to an increase in the number of DNA sequences within the genome. For example, the genomes of many tumors are now known to contain regions that have been amplified many fold compared to their non-tumor counterparts (i.e. a sequence or region of DNA that normally occurs once at a particular chromosomal location may be present in hundreds of copies in some tumors). It is believed that many such regions harbor **oncogenes**, which, when present in high copy number, predispose to development of the malignant **phenotype**.

Aneuploid

Possessing an incorrect number (abnormal complement) of **chromosomes**. The normal human complement is 46 chromosomes, any cell that deviates from this number is said to be aneuploid.

Aneuploidy

The chromosomal condition of a cell or organism with an incorrect number of **chromosomes**. Individuals with Down syndrome are described as having aneuploidy, because they possess an extra copy of chromosome 21 (**trisomy** 21), making a total of 47 chromosomes.

Anticipation

A general phenomenon that refers to the observation of an increase in severity, and/or decrease in age of onset, of a condition in successive generations of a family (see Figure 1). Anticipation is now known, in many cases, to result directly from the presence of a **dynamic mutation** in a family. In the absence of a dynamic mutation, anticipation may be explained by '**ascertainment bias**'. Thus, before the first dynamic mutations were described (in Fragile X and myotonic dystrophy), it was believed that ascertainment bias was the complete explanation for anticipation. There are two main reasons for ascertainment bias:

1. Identical **mutations** in different individuals often result in variable expressions of the associated **phenotype**. Thus, individuals within a family, all of whom harbor an identical mutation, may have variation in the severity of their condition.

2. Individuals with a severe phenotype are more likely to present to the medical profession. Moreover, such individuals are more likely to fail to reproduce (i.e. they are genetic lethals), often for social, rather than direct physical reasons.

For both reasons, it is much more likely that a mildly affected parent will be ascertained with a severely affected child, than the reverse. Therefore, the severity of a condition appears to increase through generations.

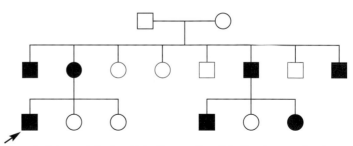

Figure 1. Autosomal dominant inheritance with **anticipation**. In many disorders that exhibit anticipation, the age of onset decreases in subsequent generations. It may happen that the transmitting parent (grandparent in this case) is unaffected at the time of presentation of the **proband** (see arrow). A good example is Huntington's disease, caused by the expansion of a CAG repeat in the coding region of the huntingtin gene. Note that this **pedigree** would also be consistent with either gonadal **mosaicism** or reduced **penetrance** (in the **carrier** grandparent).

Anticodon	The 3-base sequence on a **transfer RNA** (tRNA) molecule that is complementary to the 3-base **codon** of a **messenger RNA** (mRNA) molecule.
Ascertainment bias	See **anticipation**
Autosomal disorder	A disorder associated with a **mutation** in an autosomal **gene**.
Autosomal dominant (AD) inheritance	An **autosomal disorder** in which the **phenotype** is expressed in the **heterozygous** state. These disorders are not sex-specific. Fifty percent of offspring (when only one parent is affected) will usually manifest the disorder (see Figure 2). Marfan syndrome is a good example of an AD disorder; affected individuals possess one wild-type (normal) and one mutated **allele** at the *FBN1* **gene**,

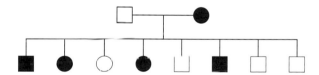

Figure 2. Autosomal dominant (AD) inheritance.

Autosomal recessive (AR) inheritance	An **autosomal disorder** in which the **phenotype** is manifest in the **homozygous** state. This pattern of inheritance is not sex-specific and is difficult to trace through generations because both parents must contribute the abnormal **gene**, but may not necessarily display the disorder. The children of two **heterozygous** AR parents have a 25% chance of manifesting the disorder (see Figure 3). Cystic fibrosis (CF) is a good example of an AR disorder; affected individuals possess two **mutations**, one at each **allele**.

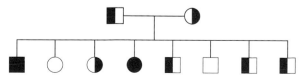

Figure 3. Autosomal recessive (AR) inheritance.

Autosome

Any **chromosome**, other than the **sex chromosomes** (X or Y), that occurs in pairs in **diploid** cells.

B

Barr body

An inactive **X chromosome**, visible in the **somatic cells** of individuals with more than one X chromosome (i.e. all normal females and all males with Klinefelter's syndrome). For individuals with n X chromosomes, n-1 Barr bodies are seen. The presence of a Barr body in cells obtained by **amniocentesis** or **chorionic villus sampling** used to be used as an indication of the sex of a baby before birth.

Base pair (bp)

Two **nucleotides** held together by hydrogen bonds. In **DNA, guanine** always pairs with **cytosine**, and **thymine** with **adenine**. A base pair is also the basic unit for measuring DNA length.

C

Carrier

An individual who is **heterozygous** for a mutant **allele** (i.e. carries one wild-type (normal copy) and one mutated copy of the **gene** under consideration).

CentiMorgan (cM)

Unit of genetic distance. If the chance of **recombination** between two loci is 1%, the loci are said to be 1 cM apart. On average, 1 cM implies a physical distance of 1 Mb (1,000,000 **base pairs**) but significant deviations from this rule of thumb occur because recombination frequencies vary throughout the **genome**. Thus if recombination in a certain region is less likely than average, 1 cM may be equivalent to 5 Mb (5,000,000 base pairs) in that region.

Centromere

Central constriction of the **chromosome** where daughter **chromatids** are joined together, separating the short (p) from the long (q) arms (see Figure 4).

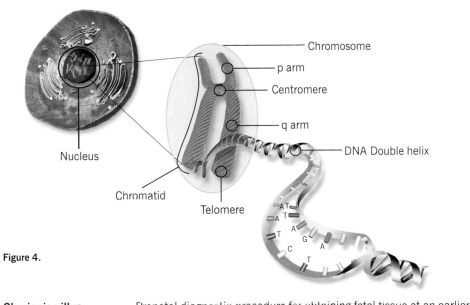

Figure 4.

Chorionic villus sampling (CVS)	Prenatal diagnostic procedure for obtaining fetal tissue at an earlier stage of gestation than **amniocentesis**. Generally performed after 10 weeks, ultrasound is used to guide aspiration of tissue from the villus area of the chorion.
Chromatid	One of the two parallel identical strands of a **chromosome**, connected at the **centromere** during **mitosis** and **meiosis** (see Figure 4). Before replication, each chromosome consists of only one chromatid. After replication, two identical sister chromatids are present. At the end of mitosis or meiosis, the two sisters separate and move to opposite poles before the cell splits.
Chromatin	A readily stained substance in the nucleus of a cell consisting of **DNA** and proteins. During cell division it coils and folds to form the metaphase **chromosomes**.
Chromosome	One of the threadlike 'packages' of **genes** and other **DNA** in the nucleus of a cell (see Figure 4). Humans have 23 pairs of chromosomes, 46 in total: 44 **autosomes** and two **sex chromosomes**. Each parent contributes one chromosome to each pair.

Chromosomal disorder	A disorder that results from gross changes in **chromosome** dose. May result from addition or loss of entire chromosomes or just portions of chromosomes.
Clone	A group of genetically identical cells with a common ancestor.
Codon	A three-base coding unit of **DNA** that specifies the function of a corresponding unit (**anticodon**) of **transfer RNA** (tRNA).
Complementary DNA (cDNA)	**DNA** synthesized from **messenger RNA** (mRNA) using **reverse transcriptase**. Differs from **genomic** DNA because it lacks **introns**.
Complementation	The wild-type **allele** of a **gene** compensates for a mutant allele of the same gene so that the heterozygote's **phenotype** is wild-type.
Complementation analysis	A genetic test (usually performed *in vitro*) that determines whether or not two **mutations** that produce the same **phenotype** are allelic. It enables the geneticist to determine how many distinct **genes** are involved when confronted with a number of mutations that have similar phenotypes.

Occasionally it can be observed clinically. Two parents who both suffer from **recessive** deafness (i.e. both are **homozygous** for a mutation resulting in deafness) may have offspring that have normal hearing. If A and B refer to the wild-type (normal) forms of the genes, and a and b the mutated forms, one parent could be aa, BB and the other AA, bb. If **alleles** A and B are distinct, each child will have the **genotype** aA, bB and will have normal hearing. If A and B are allelic, the child will be homozygous at this **locus** and will also suffer from deafness. |
| **Compound heterozygote** | An individual with two different mutant **alleles** at the same **locus**. |
| **Concordant** | A pair of twins who manifest the same **phenotype** as each other. |

Consanguinity	Sharing a common ancestor, and thus genetically related. **Recessive** disorders are seen with increased frequency in consanguineous families.
Consultand	An individual seeking genetic advice.
Contiguous gene syndrome	A syndrome resulting from the simultaneous functional imbalance of a group of **genes** (see Figure 5). The nomenclature for this group of disorders is somewhat confused, largely as a result of the history of their elucidation. The terms submicroscopic rearrangement/deletion/duplication and micro-rearrangement/deletion/duplication are often used interchangeably. Micro or submicroscopic refer to the fact that such lesions are not detectable with standard cytogenetic approaches (where the limit of resolution is usually 10 Mb, and 5 Mb in only the most fortuitous of circumstances). A newer, and perhaps more comprehensive, term that is currently applied to this group of disorders is segmental aneusomy disorders (SASs). This term embraces the possibility not only of loss or gain of a

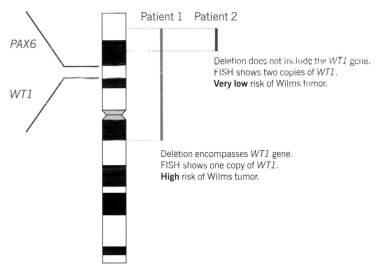

Figure 5. WAGR syndrome at 11p13. Patients 1 and 2 both have sporadic aniridia and standard **chromosome** analysis is normal (microdeletion). **FISH** analysis reveals two copies of *WT1* in patient 2 but only one in patient 1—patient 1 is thus at high risk of Wilms tumor.

chromosomal region that harbors many genes (leading to imbalance of all those genes), but also of functional imbalance in a group of genes, as a result of an abnormality of the machinery involved in their silencing/**transcription** (i.e. methylation-based mechanisms that depend on a master control gene).

In practice, most contiguous gene syndromes result from the **heterozygous** deletion of a segment of **DNA** that is large in molecular terms but not detectable cytogenetically. The size of such deletions is usually 1.5–3 Mb. It is common for one to two dozen genes to be involved in such deletions, and the resultant **phenotypes** are often complex, involving multiple organ systems and, almost invariably, learning difficulties. A good example of a contiguous gene syndrome is Williams syndrome, a sporadic disorder that is due to a heterozygous deletion at **chromosome** 7q11.23. Affected individuals have characteristic phenotypes, including recognizable facial appearance and typical behavioral traits (including moderate learning difficulties). Velocardiofacial syndrome is currently the most common **microdeletion** known, and is caused by deletions of 3 Mb at chromosome 22q11.

Crossing over Reciprocal exchange of genetic material between **homologous chromosomes** at **meiosis** (see Figure 6).

Cytogenetics The study of the structure of **chromosomes**.

Cytosine (C) One of the bases making up **DNA** and **RNA** (pairs with **guanine**).

Cytotrophoblast Cells obtained from fetal chorionic villi by chorionic villus sampling (CVS). Used for **DNA** and **chromosome** analysis.

D

Deletion A particular kind of **mutation** that involves the loss of a segment of **DNA** from a **chromosome** with subsequent re-joining of the two extant ends. It can refer to the removal of one or more bases within a **gene** or to a much larger aberration involving millions of bases. The

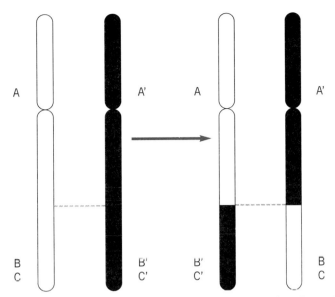

Figure 6. Schematic demonstrating the principle of **recombination (crossing over)**. On average, 50 recombinations occur per meiotic division (1–2 per **chromosome**). Loci that are far apart on the chromosome are more likely to be separated during recombination than those that are physically close to each other (they are said to be linked, see **linkage**), i.e. A and B are less likely to co-segregate than B and C. Note that the two **homologues** of a sequence have been differentially labeled according to their chromosome of origin.

term deletion is not totally specific, and differentiation must be made between **heterozygous** and **homozygous** deletions. Large heterozygous deletions are a common cause of complex **phenotypes** (see **contiguous gene syndrome**); large germ-line homozygous deletions are extremely rare but have been described. Homozygous deletions are frequently described in **somatic cells**, in association with the manifestation of the malignant phenotype. The two deletions in a homozygous deletion need not be identical but must result in the complete absence of DNA sequences that occupy the 'overlap' region.

Denature   Broadly used to describe two general phenomena:

1. The 'melting' or separation of double-stranded **DNA** (dsDNA) into its constituent single strands, which may be achieved using heat or chemical approaches.

2. The denaturation of proteins. The specificity of proteins is a result of their 3-dimensional conformation, which is a function of their (linear) amino acid sequence. Heat and/or chemical approaches may result in denaturation of a protein—the protein loses its 3-dimensional conformation (usually irreversibly) and, with it, its specific activity.

Diploid

The number of **chromosomes** in most human **somatic cells** (46). This is double the number found in **gametes** (23, the **haploid** number).

Discordant

A pair of twins who differ in their manifestation of a **phenotype**.

Dizygotic

The fertilization of 2 separate eggs by 2 separate sperm resulting in a pair of genetically non-identical twins.

DNA
(deoxyribonucleic acid)

The molecule of heredity. DNA normally exists as a double-stranded (ds) molecule; one strand is the complement (in sequence) of the other. The two strands are joined together by hydrogen bonding, a non-covalent mechanism that is easily reversible using heat or chemical means. DNA consists of 4 distinct bases: **guanine** (G), **cytosine** (C), **thymine** (T) and **adenine** (A). The convention is that DNA sequences are written in a 5' to 3' direction, where 5' and 3' refer to the numbering of carbons on the deoxyribose ring. A guanine on one strand will always pair with a cytosine on the other strand, while thymine pairs with adenine. Thus, given the sequence of bases on one strand, the sequence on the other is immediately determined:

5' – AGTGTGACTGATCTTGGTG – 3'
3' – TCACACTGACTAGAACCAC – 5'

The complexity (informational content) of a DNA molecule resides almost completely in the particular sequence of its bases. For a sequence of length 'n' **base pairs**, there are 4^n possible sequences. Even for relatively small n, this number is astronomical ($4^\mu = 1.6 \times 10^{60}$ for n = 100).

The complementarity of the two strands of a dsDNA molecule is a very important feature and one that is exploited in almost all molecular genetic techniques. If dsDNA is **denatured**, either by heat or by chemical means, the two strands become separated from each other. If the conditions are subsequently altered (e.g. by reducing heat), the two strands eventually 'find' each other in solution and re-anneal to form dsDNA once again. The specificity of this reaction is quite high, under the right circumstances—strands that are not highly complementary are much less likely to re-anneal compared to perfect or near perfect matches. The process by which the two strands 'find' each other depends on random molecular collisions, and a '**zippering**' mechanism, which is initiated from a short stretch of complementarity. This property of DNA is vital for polymerase chain reaction (PCR), **Southern blotting** and any method that relies on the use of a DNA/**RNA probe** to detect its counterpart in a complex mix of molecules.

DNA chip

A 'chip' or microarray of multiple **DNA** sequences immobilized on a solid surface (see Figure 7). The term chip refers more often to semiconductor-based DNA arrays, in which short DNA sequences (oligos) are synthesized *in situ*, using a photolithographic process akin to that used in the manufacture of semiconductor devices for the electronics industry. The term microarray is much more general and includes any collection of DNA sequences immobilized onto a solid surface, whether by a photolithographic process, or by simple 'spotting' of DNA sequences onto glass slides.

The power of DNA microarrays is based on the parallel analysis that they allow for. In conventional **hybridization** analysis (i.e. **Southern blotting**), a single DNA sequence is usually used to interrogate a small number of different individuals. In DNA microarray analysis, this approach is reversed—an individual's DNA is hybridized to an array that may contain 30,000 distinct spots. This allows for direct information to be obtained about all DNA sequences on the array in one experiment. DNA microarrays have been used successfully to directly uncover **point mutations** in single **genes**, as well as detect

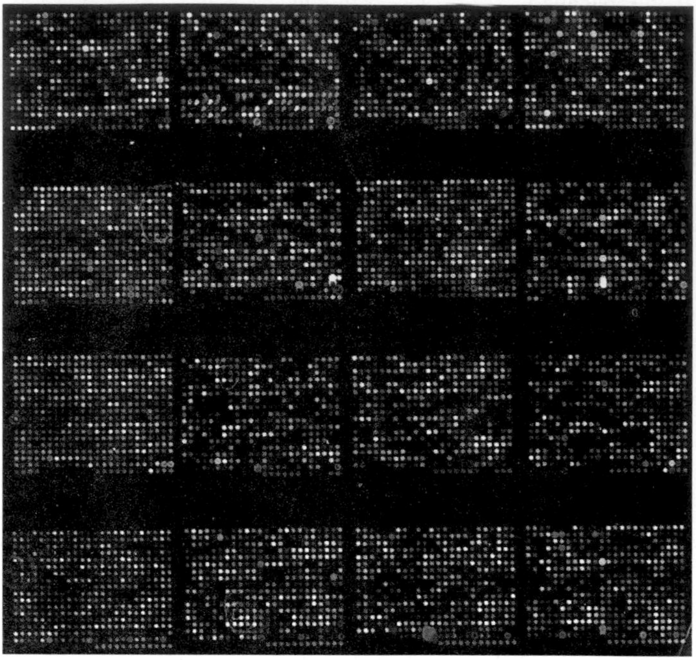

Figure 7. DNA chip. DNA arrays (or 'chips') are composed of thousands of 'spots' of DNA, attached to a solid surface (normally glass). Each spot contains a different DNA sequence. The arrays allow for massively parallel experiments to be performed on samples. In practice, two samples are applied to the array. One sample is a control (from a 'normal' sample) and one is the test sample. Each sample is labeled with fluorescent tags, control with green and test with red. The two labeled samples are co-hybridized to the array and the results read by a laser scanner. Spots on the array whose DNA content is equally represented in the test and control samples yield equal intensities in the red and green channels, resulting in a yellow signal. Spots appearing as red represent DNA sequences that are present at higher concentration in the test sample compared to the control sample and vice versa.

alterations in **gene expression** associated with certain disease states/cellular differentiation. It is likely that certain types of array will be useful in the determination of subtle copy number alterations, as occurs in **microdeletion/microduplication** syndromes.

DNA methylation   Addition of a methyl group (-CH$_3$) to **DNA nucleotides** (often **cytosine**). Methylation is often associated with reduced levels of expression of a given **gene** and is important in **imprinting**.

DNA replication Use of existing **DNA** as a template for the synthesis of new DNA strands. In humans and other eukaryotes, replication takes place in the cell nucleus. DNA replication is semi-conservative—each new double-stranded molecule is composed of a newly synthesized strand and a pre-existing strand.

Dominant (traits/diseases) Manifesting a **phenotype** in the **heterozygous** state. Individuals with Huntington's disease, a dominant condition, are affected even though they possess one normal copy of the **gene**.

Dynamic/ non-stable mutation The vast majority of **mutations** known to be associated with human genetic disease are inter-generationally stable (no alteration in the mutation is observed when transmitted from parent to child). However, a recently described and growing class of disorders result from the presence of mutations that are unstable inter-generationally. These disorders result from the presence of tandem repeats of short **DNA** sequences (e.g. the sequence CAG may be repeated many times in tandem), see Table 1. For reasons that are not completely clear, the copy number of such repeats may vary from parent to child (usually resulting in a copy number increase) and within the **somatic cells** of a given individual. Abnormal **phenotypes** result when the number of repeats reaches a given threshold. Furthermore, when this threshold has been reached, the risk of even greater expansion of copy number in subsequent generations increases.

E

Electrophoresis The separation of molecules according to size and ionic charge by an electrical current.

Agarose gel electrophoresis
Separation, based on size, of **DNA/RNA** molecules through agarose. Conventional agarose gel electrophoresis generally refers to electrophoresis carried out under standard conditions, allowing the resolution of molecules that vary in size from a few hundred to a few thousand **base pairs**.

Table 1. 'Classical' repeat expansion disorders.

Disorder	Protein/location	Repeat	Repeat location	Normal range	Pre-mutation	Full mutation	Type	MIM
Progressive myoclonus epilepsy of Unverricht-Lundborg type (EPM1)	cystatin B 21q22.3	C4GC4G CG	Promoter	2-3	12-17	30-75	AR	254800
Fragile X type A (FRAXA)	FMR1 Xq27.3	CGG	5'UTR	6-52	~60-200	~200->2,000	XLR	309550
Fragile X type E (FRAXE)	FMR2 Xq28	CGG 5	C'UTR	6-25	-	>200	XLR	309548
Friedreich's ataxia (FRDA)	frataxin 9q13	GAA	intron	17-22	-	200->900	AR	229300
Huntington's disease (HD)	huntingtin 4p16.3	CAG	ORF	6-34	-	36-180	AD	143100
Dentatorubal-pallidoluysian atrophy (DRPLA)	atrophin 12p12	CAG	ORF	7-25	-	49-88	AD	125370
Spinal and bulbar muscular atrophy (SBMA – Kennedy syndrome)	androgen receptor Xq11-12	CAG	ORF	11-24	-	40-62	XLR	313200
Spinocerebellar ataxia type 1 (SCA1)	ataxin-1 6p23	CAG	ORF	6-39	-	39-83	AD	164400
Spinocerebellar ataxia type 2 (SCA2)	ataxin-2 12q24	CAG	ORF	15-29	-	34-59	AD	183090
Spinocerebellar ataxia type 3 (SCA3)	ataxin-3 14q24.3-q31	CAG	ORF	13-36	-	55-84	AD	109150
Spinocerebellar ataxia type 6 (SCA6)	PQ calcium channel 19p13	CAG	ORF	4-16	-	21-30	AD	183086
Spinocerebellar ataxia type 7 (SCA7)	ataxin-7 3p21.1-p12	CAG	ORF	4-35	28-35	34->300	AD	164500
Spinocerebellar ataxia type 8 (SCA8)	SCA8 13q21	CTG	3'UTR	6-37	-	~107-250[1]	AD	603680
Spinocerebellar ataxia type 10 (SCA10)	SCA10 22q13-qter	ATTCT	intron 9	10-22	-	500-4,500	AD	603516
Spinocerebellar ataxia type 12 (SCA12)	PP2R2B 5q31-33	CAG	5'UTR	7-28	-	66-78	AD	604326
Myotonic dystrophy (DM)	DMPK 19q13.3	CTG	3'UTR	5-37	~50-180	~200->2,000	AD	160900

[1]Longer alleles exist but are not associated with disease.
AD: autosomal dominant; AR: autosomal recessive; ORF: open reading frame (coding region); 3' UTR: 3' untranslated region (downstream of gene); 5' UTR: 5' untranslated region (upstream of gene); XLR: X-linked recessive.

Polyacrylamide gel electrophoresis
Allows resolution of proteins or DNA molecules differing in size by only 1 base pair.

Pulsed field gel electrophoresis
(Also performed using agarose) refers to a specialist technique that allows resolution of much larger DNA molecules, in some cases up to a few Mb in size.

Empirical recurrence risk – recurrence risk   Based on observation, rather than detailed knowledge of, e.g., modes of inheritance or environmental factors.

Endonuclease   An enzyme that cleaves **DNA** at an internal site (see also **restriction enzyme**).

Euchromatin   **Chromatin** that stains lightly with trypsin G banding and contains active/potentially active **genes**.

Euploidy   Having a normal **chromosome** complement.

Exon   Coding part of a **gene**. Historically, it was believed that all of a **DNA** sequence is mirrored exactly on the messenger **RNA** (mRNA) molecule (except for the presence of **uracil** in mRNA compared to **thymine** in DNA). It was a surprise to discover that this is generally not the case. The **genomic** sequence of a gene has two components: exons and **introns**. The exons are found in both the genomic sequence and the mRNA, whereas the introns are found only in the genomic sequence. The mRNA for dystrophin, an **X-linked** gene associated with Duchenne muscular dystrophy (DMD), is 14,000 **base pairs** long but the genomic sequence is spread over a distance of 1.5 million base pairs, because of the presence of very long intronic sequences. After the genomic sequence is initially transcribed to RNA, a complex system ensures specific removal of introns. This system is known as **splicing**.

Expressivity   Degree of expression of a disease. In some disorders, individuals carrying the same **mutation** may manifest wide variability in severity of the disorder. **Autosomal dominant** disorders are often associated with **variable expressivity**, a good example being Marfan's syndrome. Variable expressivity is to be differentiated from **incomplete penetrance**, an all or none phenomenon that refers to the complete absence of a **phenotype** in some **obligate carriers**.

C A T G T T T T C C C C C A C C C A

PITX2 sequence

Mutant (protein)	ATG Met	TTT Phe	TCC Ser	CCC Pro	ACC Thr	CAA Gln
Normal (protein)	ATG Met	TTT Phe	TCC Ser	CCA Pro	CCC Pro	AAC Asn

Figure 8. Frameshift mutation. This example shows a sequence of *PITX2* in a patient with Rieger's syndrome, an **autosomal dominant** condition. The sequence graph shows only the abnormal sequence. The arrow indicated the insertion of a single **cytosine** (C) residue. When translated the triplet code is now out of frame by one base pair. This totally alters the translated protein's amino acid sequence. This leads to a premature **stop codon** later in the protein and results in Rieger's syndrome.

F

Familial

Any trait that has a higher frequency in relatives of an affected individual than the general population.

FISH

Fluorescence *in situ* hybridization (see *In situ* **hybridization**).

Founder effect

The high frequency of a mutant **allele** in a population as a result of its presence in a founder (ancestor). Founder effects are particularly noticeable in relative genetic isolates, such as the Finnish or Amish.

Frameshift mutation

Deletion/insertion of a **DNA** sequence that is not an exact multiple of 3 **base pairs**. The result is an alteration of the reading frame of the **gene** such that all sequence that lies beyond the **mutation** is effectively nonsense (see Figure 8). A premature **stop codon** is usually encountered shortly after the frameshift.

G

Gamete (germ cell)
The mature male or female reproductive cells, which contain a **haploid** set of **chromosomes**.

Gene
An ordered, specific sequence of **nucleotides** that controls the transmission and expression of one or more traits by specifying the sequence and structure of a particular protein or **RNA** molecule. Mendel defined a gene as the basic physical and functional unit of all heredity.

Gene expression
The process of converting a **genes** coded information into the existing, operating structures in the cell.

Gene mapping
Determines the relative positions of **genes** on a **DNA** molecule and plots the genetic distance in **linkage** units (**centiMorgans**) or physical distance (**base pairs**) between them.

Genetic code
Relationship between the sequence of bases in a nucleic acid and the order of amino acids in the polypeptide synthesized from it (see Table 2). A sequence of three nucleic acid bases (a triplet) acts as a codeword (**codon**) for one amino acid or instruction (start/stop).

Genetic counselling
Information/advice given to families with, or at risk of, genetic disease. Genetic counselling is a complex discipline that requires accurate diagnostic approaches, up-to-date knowledge of the genetics of the condition, an insight into the beliefs/anxieties/wishes of the individual seeking advice, intelligent risk estimation and, above all, skill in communicating relevant information to individuals from a wide variety of educational backgrounds. Genetic counselling is most often carried out by trained medical geneticists or, in some countries, specialist genetic counsellors or nurses.

Genetic heterogeneity
Association of a specific **phenotype** with **mutations** at different loci. The broader the phenotypic criteria, the greater the heterogeneity

		2nd	2nd	2nd	2nd		
		T	C	A	G		
1st	T	TTT Phe [F]	TCT Ser [S]	TAT Tyr [Y]	TGT Cys [C]	T	3rd
		TTC Phe [F]	TCC Ser [S]	TAC Tyr [Y]	TGC Cys [C]	C	
		TTA Leu [L]	TCA Ser [S]	TAA Ter [end]	TGA Ter [end]	A	
		TTG Leu [L]	TCG Ser [S]	**TAG Ter [end]**	TGG Trp [W]	G	
1st	C	CTT Leu [L]	CCT Pro [P]	CAT His [H]	CGT Arg [R]	T	3rd
		CTC Leu [L]	CCC Pro [P]	CAC His [H]	CGC Arg [R]	C	
		CTA Leu [L]	CCA Pro [P]	CAA Gln [Q]	CGA Arg [R]	A	
		CTG Leu [L]	CCG Pro [P]	CAG Gln [Q]	CGG Arg[R]	G	
1st	A	ATT Ile [I]	ACT Thr [T]	AAT Asn [N]	AGT Ser [S]	T	3rd
		ATC Ile [I]	AAC Asn [N]	ACC Thr [T]	AGC Ser [S]	C	
		ATA Ile [I]	AAA Lys [K]	ACA Thr [T]	AGA Arg [R]	A	
		ATG Met [M]	ACG Thr [T]	AAG Lys [K]	AGG Arg [R]	G	
1st	G	GTT Val [V]	GCT Ala [A]	GAT Asp [D]	GGT Gly [G]	T	3rd
		GTC Val [V]	GCC Ala [A]	GAC Asp [D]	GGC Gly [G]	C	
		GTA Val [V]	GCA Ala [A]	GAA Glu [E]	GGA Gly [G]	A	
		GTG Val [V]	GCG Ala [A]	GAG Glu [E]	GGG Gly [G]	G	

Table 2. The **genetic code**. To locate a particular codon (e.g. TAG) locate the first base (T) in the left hand column, then the second base (A) by looking at the top row, and finally the third (G) in the right hand column (TAG is a stop codon). Note the redundancy of the genetic code—for example, three different codons specify a stop signal, and threonine (Thr) is specified by any of ACT, ACC, ACA and ACG.

(e.g. mental retardation). However, even very specific phenotypes may be genetically heterogeneous. Tuberous sclerosis is a good example: this **autosomal dominant** condition is now known to be associated (in different individuals) with mutations either in the *TSC1* **gene** at 9q34 or the *TSC2* gene at 16p13.3. There is no obvious distinction between the clinical phenotypes associated with these two genes. **Genetic heterogeneity** should not be confused with **allelic heterogeneity**, which refers to the presence of different mutations at the same **locus**.

Genetic locus

A specific location on a **chromosome**.

Genetic map

A map of genetic landmarks deduced from **linkage (recombination) analysis**. Aims to determine the linear order of a set of **genetic markers** along a **chromosome**. Genetic maps differ significantly from **physical maps**, in that recombination frequencies are not identical across different **genomic** regions, resulting occasionally in large discrepancies.

Genetic marker	A **gene** that has an easily identifiable **phenotype** so that one can distinguish between those cells or individuals that do or do not have the gene. Such a gene can also be used as a **probe** to mark cell nuclei or **chromosomes**, so that they can be isolated easily or identified from other nuclei or chromosomes later.
Genetic screening	Population analysis designed to ascertain individuals at risk of either suffering or transmitting a genetic disease.
Genetically lethal	Prevents reproduction of the individual, either because the condition causes death prior to reproductive age, or because social factors make it highly unlikely (although not impossible) that the individual concerned will reproduce.
Genome	The complete **DNA** sequence of an individual, including the **sex chromosomes** and **mitochondrial DNA**. The genome of humans is estimated to have a complexity of 3.3×10^9 **base pairs** (per **haploid** genome).
Genomic	Pertaining to the **genome**. Genomic **DNA** differs from **complementary DNA** in that it contains non-coding as well as coding DNA.
Genotype	Genetic constitution of an individual, distinct from expressed features (**phenotype**).
Germ line	Germ cells (those cells that produce **haploid gametes**) and the cells from which they arise. The germ line is formed very early in embryonic development. Germ line **mutations** are those present constitutionally in an individual (i.e. in all cells of the body) as opposed to somatic mutations, which affect only a proportion of cells.
Giemsa banding	Light/dark bar code obtained by staining **chromosomes** with Giemsa stain. Results in a unique bar code for each chromosome.
Guanine (G)	One of the bases making up **DNA** and **RNA** (pairs with **cytosine**).

H

Haploid The **chromosome** number of a normal **gamete**, containing one each of every individual chromosome (23 in humans).

Haploinsufficiency The presence of one active copy of a **gene**/region is insufficient to compensate for the absence of the other copy. Most genes are not 'haploinsufficient'—50% reduction of gene activity does not lead to an abnormal **phenotype**. However, for some genes, most often those involved in early development, reduction to 50% often correlates with an abnormal phenotype. Haploinsufficiency is an important component of most **contiguous gene disorders** (e.g. in Williams syndrome, **heterozygous deletion** of a number of genes results in the mutant phenotype, despite the presence of normal copies of all affected genes).

Hemizygous Having only one copy of a **gene** or **DNA** sequence in **diploid** cells. Males are hemizygous for most genes on the **sex chromosomes**, as they possess only one **X chromosome** and one **Y chromosome** (the exceptions being those genes with counterparts on both sex chromosomes). **Deletions** on **autosomes** produce hemizygosity in both males and females.

Heterochromatin Contains few active **genes**, but is rich in highly repeated simple sequence **DNA**, sometimes known as satellite DNA. Heterochromatin refers to inactive regions of the **genome**, as opposed to **euchromatin**, which refers to active, gene expressing regions. Heterochromatin stains darkly with Giemsa.

Heterozygous Presence of two different **alleles** at a given **locus**.

Histones Simple proteins bound to **DNA** in **chromosomes**. They help to maintain **chromatin** structure and play an important role in regulating **gene** expression.

Holandric

Pattern of inheritance displayed by **mutations** in **genes** located only on the **Y chromosome**. Such mutations are transmitted only from father to son.

Homologue or homologous gene

Two or more **genes** whose sequences manifest significant similarity because of a close evolutionary relationship. May be between species (orthologues) or within a species (paralogues).

Homologous chromosomes

Chromosomes that pair during **meiosis**. These chromosomes contain the same linear **gene** sequences as one another and derive from one parent.

Homology

Similarity in **DNA** or protein sequences between individuals of the same species or among different species.

Homozygous

Presence of identical **alleles** at a given **locus**.

Human gene therapy

The study of approaches to treatment of human genetic disease, using the methods of modern molecular genetics. Many trials are underway studying a variety of disorders including cystic fibrosis. Some disorders are likely to be more treatable than others—it is probably going to be easier to replace defective or absent **gene** sequences rather than deal with genes whose aberrant expression results in an actively toxic effect.

Human genome project

Worldwide collaboration aimed at obtaining a complete sequence of the human **genome**. Most sequencing has been carried out in the USA, although the Sanger Centre in Cambridge, UK has sequenced one third of the genome, and centers in Japan and Europe have also contributed significantly. The first draft of the human genome was released in the summer of 2000 to much acclaim. The finished sequence may not be available until 2003. Celera, a privately funded venture, headed by Dr Craig Ventner, also published its first draft at the same time.

| **Hybridization** | Pairing of complementary strands of nucleic acid. Also known as **re-annealing**. May refer to re-annealing of **DNA** in solution, on a membrane (**Southern blotting**) or on a DNA microarray. May also be used to refer to fusion of two **somatic cells**, resulting in a hybrid that contains genetic information from both donors. |

I

| **Imprinting** | A general term used to describe the phenomenon whereby a **DNA** sequence (coding or otherwise) carries a signal or imprint that indicates its parent of origin. For most DNA sequences, no distinction can be made between those arising paternally and those arising maternally (apart from subtle sequence variations); for imprinted sequences this is not the case. The mechanistic basis of imprinting is almost always methylation—for certain **genes**, the copy that has been inherited from the father is methylated, while the maternal copy is not. The situation may be reversed for other imprinted genes. Note that imprinting of a gene refers to the general phenomenon, not which parental copy is methylated (and, therefore, usually inactive). Thus, formally speaking, it is incorrect to say that a gene undergoes paternal imprinting. It is correct to say that the gene undergoes imprinting and that the inactive (methylated) copy is always the paternal one. However, in common genetics parlance, paternal imprinting is usually understood to mean the same thing. |

| ***In situ* hybridization** | Annealing of **DNA** sequences to immobilized **chromosomes**/cells/tissues. Historically done using radioactively labeled **probes**, this is currently most often performed with fluorescently tagged molecules (fluorescent *in situ* hybridization – **FISH**, see Figure 9). ISH/FISH allows for the rapid detection of a DNA sequence within the **genome**. |

| **Incomplete penetrance** | Complete absence of expression of the abnormal **phenotype** in a proportion of individuals known to be **obligate carriers**. To be distinguished from **variable expressivity**, in which the phenotype always manifests in obligate carriers but with widely varying degrees of severity. |

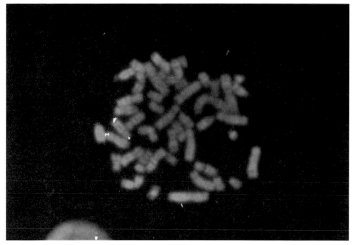

Figure 9. Fluorescence *in situ* hybridization. FISH analysis of a patient with a complex syndrome, using a clone containing **DNA** from the region 8q24.3. In addition to that clone, a control from 8pter was used. The 8pter clone has yielded a signal on both **homologues** of **chromosome** 8, while the 'test' clone from 8q24.3 has yielded a signal on only one homologue, demonstrating a (**heterozygous**) deletion in that region.

Index case – proband

The individual through which a family medically comes to light. For example, the index case may be a baby with Down syndrome. Can be termed propositus (if male) or proposita (if female).

Insertion

Interruption of a chromosomal sequence as a result of insertion of material from elsewhere in the **genome** (either a different **chromosome**, or elsewhere from the same chromosome). Such insertions may result in abnormal **phenotypes** either because of direct interruption of a **gene** (uncommon), or because of the resulting imbalance (i.e. increased dosage) when the chromosomes that contain the normal counterparts of the inserted sequence are also present.

Intron

A non-coding **DNA** sequence that 'interrupts' the protein-coding sequences of a **gene**; intron sequences are transcribed into **messenger RNA** (mRNA) but are cut out before the mRNA is translated into a protein (this process is known as **splicing**). Introns may contain sequences involved in regulating expression of a gene. Unlike the

exon, the intron is the **nucleotide** sequence in a gene that is not represented in the amino acid sequence of the final gene product.

Inversion

A structural abnormality of a **chromosome** in which a segment is reversed, as compared to the normal orientation of the segment. An inversion may result in the reversal of a segment that lies entirely on one chromosome arm (paracentric) or one that spans (i.e. contains) The **centromere** (pericentric). While individuals who possess an inversion are likely to be genetically balanced (and therefore usually phenotypically normal), they are at increased risk of producing unbalanced offspring because of problems at **meiosis** with pairing of the inversion chromosome with its normal **homologue**. Both **deletions** and duplications may result, with concomitant congenital abnormalities related to **genomic** imbalance, or miscarriage if the imbalance is lethal.

K

Karyotype

A photomicrograph of an individual's **chromosomes** arranged in a standard format showing the number, size, and shape of each chromosome type, and any abnormalities of chromosome number or morphology (see Figure 10).

Kilobase (kb)

1000 **base pairs** of **DNA**.

Knudson hypothesis

See **tumor suppressor gene**

L

Linkage

Co-inheritance of **DNA** sequences/**phenotypes** as a result of physical proximity on a **chromosome**. Before the advent of molecular genetics, linkage was often studied with regard to proteins, enzymes or cellular characteristics. An early study demonstrated linkage between the Duffy blood group and a form of **autosomal dominant** congenital cataract (both are now known to reside at 1q21.1). Phenotypes may

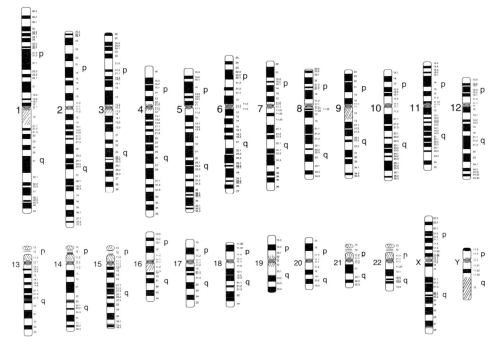

Figure 10. Schematic of a normal human (male) **karyotype**. (ISCN 550 ideogram produced by the MRC Human Genetics Unit, Edinburgh, reproduced with permission.)

also be linked in this manner (i.e. families manifesting two distinct Mendelian disorders).

During the **recombination** phase of **meiosis**, genetic material is exchanged (equally) between two **homologous chromosomes**. Genes/DNA sequences that are located physically close to each other are unlikely to be separated during recombination. Sequences that lie far apart on the same chromosome are more likely to be separated. For sequences that reside on different chromosomes, segregation will always be random, so that there will be a 50% chance of 2 markers being co-inherited.

Linkage analysis   An algorithm designed to map (i.e. physically locate) an unknown **gene** (associated with the **phenotype** of interest) to a chromosomal region. Linkage analysis has been the mainstay of disease-associated gene identification for some years. The general availability of large numbers of DNA markers that are variable in the population (**polymorphisms**), and which therefore permit **allele** discrimination, has made linkage analysis a relatively rapid and dependable approach (see Figure 11). However, the method relies on the ascertainment of large families manifesting Mendelian disorders. Relatively little phenotypic heterogeneity is tolerated, as a single misassigned individual (believed to be unaffected despite being a gene **carrier**) in a **pedigree** may completely invalidate the results. **Genetic heterogeneity** is another problem, not within families (usually) but between families. Thus, conditions that result in identical phenotypes despite being associated with **mutations** within different genes (e.g. tuberous sclerosis) are often hard to study. Linkage analysis typically follows a standard algorithm:

1. Large families with a given disorder are ascertained. Detailed clinical evaluation results in assignment of affected vs. unaffected individuals.

2. Large numbers of polymorphic DNA markers that span the **genome** are analyzed in all individuals (affected and unaffected).

3. The results are analyzed statistically, in the hope that one of the markers used will have demonstrably been co-inherited with the phenotype in question more often than would be predicted by chance.

The LOD score (**logarithm of the odds**) gives an indication of the likelihood of the result being significant (and not having occurred simply as a result of chance co-inheritance of the given marker with the condition).

Linkage disequilibrium   Association of particular **DNA** sequences with each other, more often than is likely by chance alone (see Figure 12). Of particular relevance to inbred populations (e.g. Finland), where specific disease **mutations** are found to reside in close proximity to specific variants of DNA markers, as a result of the **founder effect**.

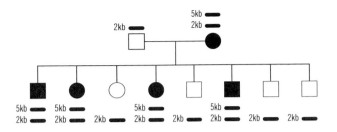

In the example above, note that the (affected) mother has a 5 kb band in addition to a 2 kb band. All the unaffected individuals have the small band only, all those who are affected have the large band. The unaffected individuals must have the mother's 2 kb fragment rather than her 5 kb fragment, and the affected individuals must have inherited the 5 kb band from the mother (as the father does not have one)—note that those individuals who only show the 2 kb band still have two alleles (one from each parent), they are just the same size and so cannot be differentiated. Thus, it appears that the 5 kb band is segregating with the disorder. The results in a family such as this are suggestive but further similar results in other families would be required for a sufficiently high **LOD** score.

The **probe** recognizes a **DNA** sequence adjacent to a restriction site (see arrow) that is polymorphic (present on some **chromosomes** but not others). When such a site is present, the **DNA** is cleaved at that point and the probe detects a 2 kb fragment. When absent, the DNA is not cleaved and the probe detects a fragment of size (2 + 3) kb = 5 kb. X refers to the points at which the **restriction enzyme** will cleave the DNA. The recognition sequence for most restriction enzymes is very stringent—change in just one **nucleotide** will result in failure of cleavage. Most RFLPs result from the presence of a single nucleotide polymorphism that has altered the restriction site.

Figure 11. Schematic demonstrating the use of **restriction fragment length polymorphisms** (RFLPs) in **linkage analysis**.

Linkage map
A map of **genetic markers** as determined by genetic analysis (i.e. **recombination** analysis). May differ markedly from a map determined by actual physical relationships of genetic markers, because of the variability of recombination.

Locus
The position of a **gene/DNA** sequence on the **genetic map**. Allelic genes/sequences are situated at identical loci in **homologous chromosomes**.

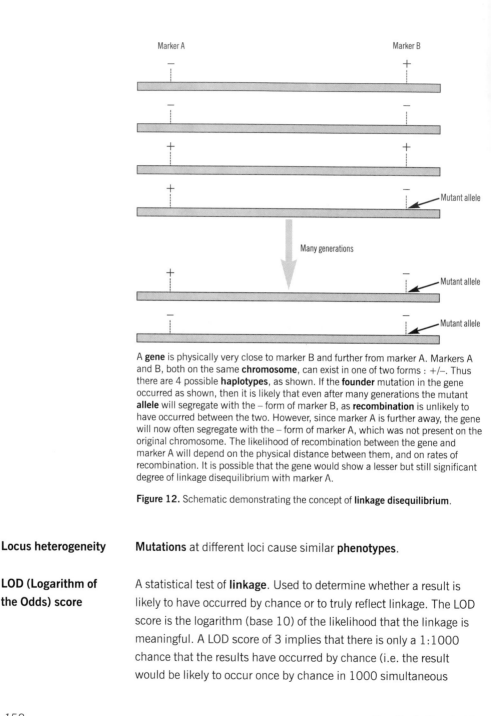

A **gene** is physically very close to marker B and further from marker A. Markers A and B, both on the same **chromosome**, can exist in one of two forms : +/−. Thus there are 4 possible **haplotypes**, as shown. If the **founder** mutation in the gene occurred as shown, then it is likely that even after many generations the mutant **allele** will segregate with the − form of marker B, as **recombination** is unlikely to have occurred between the two. However, since marker A is further away, the gene will now often segregate with the − form of marker A, which was not present on the original chromosome. The likelihood of recombination between the gene and marker A will depend on the physical distance between them, and on rates of recombination. It is possible that the gene would show a lesser but still significant degree of linkage disequilibrium with marker A.

Figure 12. Schematic demonstrating the concept of **linkage disequilibrium**.

Locus heterogeneity	Mutations at different loci cause similar phenotypes.
LOD (Logarithm of the Odds) score	A statistical test of **linkage**. Used to determine whether a result is likely to have occurred by chance or to truly reflect linkage. The LOD score is the logarithm (base 10) of the likelihood that the linkage is meaningful. A LOD score of 3 implies that there is only a 1:1000 chance that the results have occurred by chance (i.e. the result would be likely to occur once by chance in 1000 simultaneous

studies addressing the same question). This is taken as proof of linkage (see Figure 11).

Lyonisation   The inactivation of n-1 **X chromosomes** on a random basis in an individual with n X chromosomes. Named after Mary Lyon, this mechanism ensures dosage compensation of **genes** encoded by the X chromosome. X chromosome inactivation does not occur in normal males who possess only one X chromosome but does occur in one of the two X chromosomes of normal females. In males who possess more than one X chromosome (i.e. XXY, XXXY, etc.), the rule is the same and only one X chromosome remains active. X-inactivation occurs in early embryonic development and is random in each cell. The inactivation pattern in each cell is faithfully maintained in all daughter cells. Therefore, females are genetic **mosaics**, in that they possess two populations of cells with respect to the X chromosome: one population has one X active, while in the other population the other X is active. This is relevant to the expression of **X-linked** disease in females.

M

Meiosis   The process of cell division by which male and female **gametes** (germ cells) are produced. Meiosis has two main roles. The first is **recombination** (during meiosis I). The second is reduction division. Human beings have 46 **chromosomes**, and each is conceived as a result of the union of two germ cells; therefore, it is reasonable to suppose that each germ cell will contain only 23 chromosomes (i.e. the **haploid** number). If not, then the first generation would have 92 chromosomes, the second 184, etc. Thus, at meiosis I, the number of chromosomes is reduced from 46 to 23.

Mendelian inheritance   Refers to a particular pattern of inheritance, obeying simple rules: each **somatic cell** contains 2 **genes** for every characteristic and each pair of genes divides independently of all other pairs at **meiosis**.

Mendelian Inheritance in Man (MIM/OMIM)	A catalogue of human Mendelian disorders, initiated in book form by Dr Victor McKusick of Johns Hopkins Hospital in Baltimore, USA. The original catalogue (produced in the mid-1960s) listed approximately 1500 conditions. By December 1998, this number had risen to 10,000, at the time of writing (November 2001) the figure had reached 13,118. With the advent of the Internet, MIM is now available as an online resource, free of charge (OMIM – Online Mendelian Inheritance in Man). The URL for this site is: http://www.ncbi.nlm.nih.gov/omim/. The online version is updated regularly, far faster than is possible for the print version, therefore, new **gene** discoveries are quickly assimilated into the database. OMIM lists disorders according to their mode of inheritance:

1 ---- (100000-) **Autosomal dominant** (entries created before May 15, 1994)

2 ---- (200000-) **Autosomal recessive** (entries created before May 15, 1994)

3 ---- (300000-) **X-linked** loci or **phenotypes**

4 ---- (400000-) Y-linked loci or phenotypes

5 ---- (500000-) Mitochondrial loci or phenotypes

6 ---- (600000-) Autosomal loci/phenotypes (entries created after May 15, 1994).

Full explanations of the best way to search the catalogue are available at the home page for OMIM.

Messenger RNA (mRNA)	The template for protein synthesis, carries genetic information from the nucleus to the ribosomes where the code is translated into protein. Genetic information flows: **DNA** → RNA → protein.
Methylation	See **DNA methylation**
Microdeletion	Structural **chromosome** abnormality involving the loss of a segment that is not detectable using conventional (even high resolution)

cytogenetic analysis. Microdeletions usually involve 1–3 Mb of sequence (the resolution of cytogenetic analysis rarely is better than 10 Mb). Most microdeletions are **heterozygous**, although some individuals/families have been described with **homozygous** microdeletions. See also **contiguous gene syndrome**.

Microduplication   Structural **chromosome** abnormality involving the gain of a segment that may involve long sequences (commonly 1–3 Mb), which are, nevertheless, undetectable using conventional cytogenetic analysis. Patients with microduplications have 3 copies of all sequences within the duplicated segment, as compared to 2 copies in normal individuals. See also **contiguous gene syndrome**.

Microsatellites   **DNA** sequences composed of short tandem repeats (STRs), such as di- and trinucleotide repeats, distributed widely throughout the **genome** with varying numbers of copies of the repeating units. Microsatellites are very valuable as **genetic markers** for mapping human **genes**.

Missense mutation   Single base substitution resulting in a **codon** that specifies a different amino acid than the wild-type.

Mitochondrial disease/disorder   Ambiguous term referring to disorders resulting from abnormalities of mitochondrial function. Two separate possibilities should be considered.

1. **Mutations** in the mitochondrial **genome** (see Figure 13). Such disorders will manifest an inheritance pattern that mirrors the manner in which mitochondria are inherited. Therefore, a mother will transmit a mitochondrial mutation to all her offspring (all of whom will be affected, albeit to a variable degree). A father will not transmit the disorder to any of his offspring.

2. Mutations in nuclear encoded **genes** that adversely affect mitochondrial function. The mitochondrial genome does not code for all the genes required for its maintenance, many are encoded in the

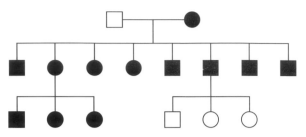

Figure 13. Mitochondrial inheritance. This **pedigree** relates to mutations in the mitochondrial **genome**.

nuclear genome. However, the inheritance patterns will differ markedly from the category described in the first option, and will be indistinguishable from standard Mendelian disorders.

Each mitochondrion possesses between 2–10 copies of its genome, and there are approximately 100 mitochondria in each cell. Therefore, each cell possesses 200–1000 copies of the mitochondrial genome. Heteroplasmy refers to the variability in sequence of this large number of genomes—even individuals with mitochondrial genome mutations are likely to have wild-type **alleles**. Variability in the proportion of molecules that are wild-type may have some bearing on the clinical variability often seen in such disorders.

Mitochondrial DNA   The **DNA** in the circular **chromosome** of mitochondria. Mitochondrial DNA is present in multiple copies per cell and mutates more rapidly than **genomic** (nuclear) DNA.

Mitosis   Cell division occurring in **somatic cells**, resulting in two daughter cells that are genetically identical to the parent cell.

Monogenic trait   Causally associated with a single **gene** (see Mendelian trait).

Monosomy   Absence of one of a pair of **chromosomes**.

Monozygotic   Arising from a single **zygote** or fertilized egg. Monozygotic twins are genetically identical.

Mosaicism or mosaic Refers to the presence of two or more distinct cell lines, all derived from the same **zygote**. Such cell lines differ from each other as a result of **DNA** content/sequence. Mosaicism arises when the genetic alteration occurs post-fertilization (post-zygotic). The important features that need to be considered in mosaicism are:

The proportion of cells that are 'abnormal'. In general, the greater proportion of cells that are abnormal, the greater the severity of the associated **phenotype**.

The specific tissues that contain high levels of the abnormal cell line(s). This variable will clearly also be relevant to the manifestation of any phenotype. An individual may have a **mutation** bearing cell line in a tissue where the mutation is largely irrelevant to the normal functioning of that tissue, with a concomitant reduction in phenotypic sequelae.

Mosaicism may be functional, as in normal females who are mosaic for activity of the two **X chromosomes** (see **Lyonisation**).

Mosaicism may occasionally be observed directly. **X-linked** skin disorders, such as incontinentia pigmenti, often manifest **mosaic** changes in the skin of a female, such that abnormal skin is observed alternately with normal skin, often in streaks (Blaschko's lines), which delineate developmental histories of cells.

Multifactorial inheritance A type of hereditary pattern resulting from a complex interplay of genetic and environmental factors.

Mutation Any heritable change in **DNA** sequence.

N

Non-disjunction Failure of two **homologous chromosomes** to pull apart during **meiosis** I, or two **chromatids** of a chromosome to separate in meiosis II or **mitosis**. The result is that both are transmitted to one daughter cell, while the other daughter cell receives neither.

Non-dynamic (stable) mutations	Stably inherited **mutations**, in contradistinction to **dynamic mutations**, which display variability from generation to generation. Includes all types of stable mutation (single base substitution, small **deletions/insertions**, **microduplications** and **microdeletions**).
Non-penetrance	Failure of expression of a **phenotype** in the presence of the relevant **genotype**.
Nonsense mutation	A single base substitution resulting in the creation of a **stop codon** (see Figure 14).
Northern blot	**Hybridization** of a radio-labeled **RNA/DNA probe** to an immobilized RNA sequence. So called in order to differentiate it from **Southern blotting** which was described first. Neither has any relationship to points on the compass. Southern blotting was named after its inventor Ed Southern (currently Professor of Biochemistry at Oxford University, UK).
Nucleotide	A basic unit of **DNA** or **RNA** consisting of a nitrogenous base— **adenine**, **guanine**, **thymine** or **cytosine** in DNA, and adenine, guanine, **uracil** or cytosine in RNA. A nucleotide is composed of a phosphate molecule, and a sugar molecule—deoxyribose in DNA and ribose in RNA. Many thousands or millions of nucleotides link to form a DNA or RNA molecule.

O

Obligate carrier	See **obligate heterozygote**
Obligate heterozygote (obligate carrier)	An individual who, on the basis of **pedigree** analysis, must carry the mutant **allele**.
Oncogene	A **gene** that, when over expressed, causes neoplasia. In contrast to **tumor suppressor genes**, which result in tumorigenesis when their activity is reduced.

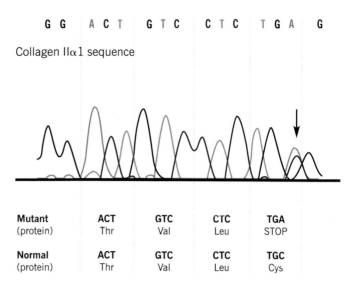

G G   A C T   G T C   C T C   T G A   G

Collagen IIα1 sequence

Mutant (protein)	ACT Thr	GTC Val	CTC Leu	TGA STOP
Normal (protein)	ACT Thr	GTC Val	CTC Leu	TGC Cys

Figure 14. Nonsense mutation. This example shows a sequence graph of collagen II (alpha 1) in a patient with Stickler syndrome, an **autosomal dominant** condition. The sequence is of **genomic DNA** and shows both normal and abnormal sequences (the patient is heterozygous for the mutation). The base marked with an arrow has been changed from C to A. When translated the codon is changed from TGC (cysteine) to TGA (stop). The premature **stop codon** in the collagen gene results in Stickler syndrome.

P

p	Short arm of a **chromosome** (from the French *petit*) (see Figure 4).
Palindromic sequence	A **DNA** sequence that contains the same 5' to 3' sequence on both strands. Most **restriction enzymes** recognize palindromic sequences. An example is 5' – AGATCT – 3', which would read 3' – TCTAGA – 5' on the complementary strand. This is the recognition site of *BglII*.
Pedigree	A schematic for a family indicating relationships to the **proband** and how a particular disease or trait has been inherited (see Figure 15).
Penetrance	An all-or-none phenomenon related to the proportion of individuals with the relevant **genotype** for a disease who actually manifest

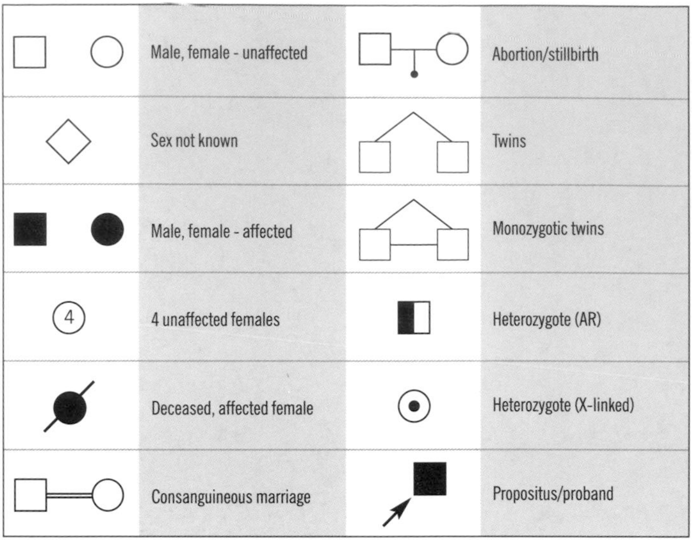

Figure 15. Symbols commonly used in **pedigree** drawing.

the **phenotype**. Note the difference between penetrance and **variable expressivity**.

Phenotype

Observed disease/abnormality/trait. An all-embracing term that does not necessarily imply pathology. A particular phenotype may be the result of **genotype**, the environment or both.

Physical map

A map of the locations of identifiable landmarks on **DNA**, such as specific DNA sequences or **genes**, where distance is measured in **base pairs**. For any **genome**, the highest resolution map is the complete **nucleotide** sequence of the **chromosomes**. A physical map should be distinguished from a **genetic map**, which depends on **recombination** frequencies.

Plasmid

Found largely in bacterial and protozoan cells, plasmids are autonomously replicating, extrachromosomal, circular **DNA** molecules that are distinct from the normal bacterial **genome** and are often used as vectors in recombinant DNA technologies. They are not essential for cell survival under non-selective conditions, but can be incorporated into the genome and are transferred between cells if they encode a protein that would enhance survival under selective conditions (e.g. an enzyme that breaks down a specific antibiotic).

Pleiotropy

Diverse effects of a single **gene** on many organ systems (e.g. the **mutation** in Marfan's syndrome results in lens dislocation, aortic root dilatation and other pathologies).

Ploidy

The number of sets of **chromosomes** in a cell. Human cells may be **haploid** (23 chromosomes, as in mature sperm or ova), **diploid** (46 chromosomes, seen in normal **somatic cells**) or triploid (69 chromosomes, seen in abnormal somatic cells, which results in severe congenital abnormalities).

Point mutation

Single base substitution.

Polygenic disease

Disease (or trait) that results from the simultaneous interaction of multiple **gene** mutations, each of which contributes to the eventual **phenotype**. Generally, each **mutation** in isolation is likely to have a relatively minor effect on the phenotype. Such disorders are not inherited in a Mendelian fashion. Examples include hypertension, obesity and diabetes.

Polymerase chain reaction (PCR)

A molecular technique for amplifying **DNA** sequences *in vitro* (see Figure 16). The DNA to be copied is **denatured** to its single strand form and two synthetic oligonucleotide primers are annealed to complementary regions of the target DNA in the presence of excess deoxynucleotides and a heat-stable DNA polymerase. The power of PCR lies in the exponential nature of **amplification**, which results from repeated cycling of the 'copying' process. Thus, a single molecule will be copied in the first cycle, resulting in 2 molecules. In the second cycle, each of these will also be copied, resulting in 4 copies. In theory, after n cycles, there will be 2^n molecules for each

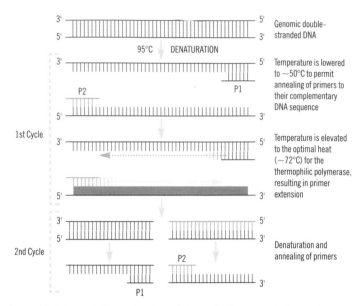

Figure 16. Schematic illustrating the technique of **polymerase chain reaction** (PCR).

starting molecule. In practice, this theoretical limit is rarely reached, mainly for technical reasons. PCR has become a standard technique in molecular biology research as well as routine diagnostics.

Polymorphism

May be applied to **phenotype** or **genotype**. The presence in a population of two or more distinct variants, such that the frequency of the rarest is at least 1% (more than can be explained by recurrent **mutation** alone). A **genetic locus** is polymorphic if its sequence exists in at least two forms in the population.

Premutation

Any **DNA mutation** that has little, if any, phenotypic consequence but predisposes future generations to the development of full mutations with phenotypic sequelae. Particularly relevant in the analysis of diseases associated with **dynamic mutations**.

Proband (propositus) – index case

The first individual to present with a disorder through which a **pedigree** can be ascertained.

Probe	General term for a molecule used to make a measurement. In molecular genetics, a probe is a piece of **DNA** or **RNA** that is labeled and used to detect its complementary sequence (e.g. **Southern blotting**).
Promoter region	The non-coding sequence upstream (5') of a **gene** where **RNA** polymerase binds. **Gene expression** is controlled by the promoter region both in terms of level and tissue specificity.
Protease	An enzyme that digests other proteins by cleaving them into small fragments; proteases may have broad specificity or only cleave a particular site on a protein or set of proteins.
Protease inhibitor	A chemical that can inhibit the activity of a **protease**. Most proteases have a corresponding specific protease inhibitor.
Proto-oncogene	A misleading term that refers to **genes** that are usually involved in signaling and cell development, and are often expressed in actively dividing cells. Certain **mutations** in such genes may result in malignant transformation, with the mutated genes being described as **oncogenes**. The term proto-oncogene is misleading because it implies that such genes were selected for by evolution in order that, upon mutation, cancers would result because of oncogenic activation. A similar problem arises with the term **tumor suppressor gene**.
Pseudogene	Near copies of true **genes**. Pseudogenes share sequence **homology** with true genes but are inactive as a result of multiple **mutations** over a long period of time.
Purine	A nitrogen-containing, double-ring, basic compound occurring in nucleic acids. The purines in **DNA** and **RNA** are **adenine** and **guanine**.
Pyrimidine	A nitrogen-containing, single-ring, basic compound that occurs in nucleic acids. The pyrimidines in **DNA** are **cytosine** and **thymine**, and cytosine and **uracil** in **RNA**.

Q

q   Long arm of a **chromosome** (see Figure 4).

R

Re-annealing   see **hybridization**

Recessive   Manifest only in homozygotes. For the **X chromosome**, recessivity
(traits, diseases)   applies to males who carry only one (mutant) **allele**. Females
                         who carry **X-linked mutations** are generally heterozygotes and,
                         barring unfortunate X-inactivation, do not manifest X-linked
                         recessive **phenotypes**.

Reciprocal translocation The exchange of material between two non-**homologous
                              chromosomes**.

Recombination   The creation of new combinations of linked **genes** as a result of
                         crossing over at **meiosis** (see Figure 6).

Recurrence risk   The chance that a genetic disease, already present in a member of a
                         family, will recur in that family and affect another individual.

Restriction enzyme   **Endonuclease** that cleaves double-stranded (ds) **DNA** at specific
                         sequences. For example, the enzyme *BglII* recognizes the sequence
                         AGATCT, and cleaves after the first A on both strands. Most
                         restriction endonucleases recognize sequences that are palindromic—
                         the complementary sequence to AGATCT, read in the same orientation,
                         is also AGATCT. The term 'restriction' refers to the function of these
                         enzymes in nature. The organism that synthesizes a given restriction
                         enzyme (e.g. *BglII*) does so in order to 'kill' foreign DNA—'restricting'
                         the potential of foreign DNA that has become integrated to adversely
                         G enzyme by simultaneously synthesizing a specific methylase that
                         recognizes the same sequence and modifies one of the bases, such

that the restriction enzyme is no longer able to cleave. Thus, for every restriction enzyme, it is likely that a corresponding methylase exists, although in practice only a relatively small number of these have been isolated.

Restriction fragment length polymorphism (RFLP)   A restriction fragment is the length of **DNA** generated when DNA is Cleaved by a **restriction enzyme**. Restriction fragment length varies when a **mutation** occurs within a restriction enzyme sequence. Most commonly the **polymorphism** is a single base substitution but it may also be a variation in length of a DNA sequence due to variable number tandem repeats (VNTRs). The analysis of the fragment lengths after DNA is cut by restriction enzymes is a valuable tool for establishing **familial** relationships and is often used in forensic analysis of blood, hair or semen (see Figure 11).

Restriction map   A **DNA** sequence map, indicating the position of restriction sites.

Reverse genetics   Identification of the causative **gene** for a disorder, based purely on molecular genetic techniques, when no knowledge of the function of the gene exists (the case for most genetic disorders).

Reverse transcriptase   Catalyses the synthesis of **DNA** from a single-stranded **RNA** template. Contradicted the central dogma of genetics (DNA → RNA → protein) and earned its discoverers the Nobel Prize in 1975.

RNA (ribonucleic acid)   RNA molecules differ from **DNA** molecules in that they contain a ribose sugar instead of deoxyribose. There are a variety of types of RNA (including **messenger RNA**, **transfer RNA** and ribosomal RNA) and they work together to transfer information from DNA to the protein-forming units of the cell.

Robertsonian translocation   A **translocation** between two acrocentric **chromosomes**, resulting from centric fusion. The short arms and satellites (chromosome segments separated from the main body of the chromosome by a constriction and containing highly repetitive **DNA**) are lost.

S

Second hit hypothesis   See **tumor suppressor gene**

Sex chromosomes   Refers to the **X** and **Y chromosomes**. All normal individuals possess 46 chromosomes, of which 44 are **autosomes** and 2 are sex chromosomes. An individual's sex is determined by his/her complement of sex chromosomes. Essentially, the presence of a Y chromosome results in the male **phenotype**. Males have an X and a Y chromosome, while females possess two X chromosomes. The Y chromosome is small and contains relatively few **genes**, concerned almost exclusively with sex determination and/or sperm formation. By contrast, the X chromosome is a large chromosome that possesses many hundreds of genes.

Sex-limited trait   A trait/disorder that is almost exclusively limited to one sex and often results from **mutations** in autosomal **genes**. A good example of a sex-limited trait is breast cancer. While males are affected by breast cancer, it is much less common (~1%) than in women. Females are more prone to breast cancer than males not only because they possess significantly more breast tissue but also because their hormonal milieu is significantly different. In many cases, early onset bilateral breast cancer is associated with mutations either in *BRCA1* or *BRCA2*, both autosomal genes. An example of a sex-limited trait in males is male pattern baldness, which is extremely rare in pre-menopausal women. The inheritance of male pattern baldness is consistent with **autosomal dominant**, not **sex-linked dominant**, inheritance.

Sex-linked dominant   See **X-linked dominant**

Sex-linked recessive   See **X-linked recessive**

Sibship   All the sibs in a family.

Silent mutation   One that has no (apparent) phenotypic effect.

Single gene disorder	A disorder resulting from a **mutation** on one **gene**.
Somatic cell	Any cell of a multicellular organism not involved in the production of **gametes**.
Southern blot	**Hybridization** with a radio-labeled **RNA/DNA probe** to an immobilized DNA sequence (see Figure 17). Named after Ed Southern (currently Professor of Biochemistry at Oxford University, UK), the technique has spawned the nomenclature for other types of blot (**Northern blots** for RNA and **Western blots** for proteins).
Splicing	Removal of **introns** from precursor **RNA** to produce **messenger RNA**. The process involves recognition of intron-**exon** junctions and specific removal of intronic sequences, coupled with re-connection of the two strands of **DNA** that formerly flanked the intron.

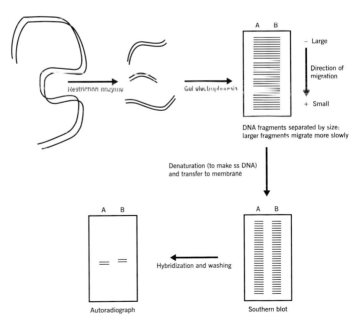

Figure 17. Southern blotting.

Start codon   The AUG **codon** of **messenger RNA** recognized by the ribosome to begin protein production.

Stop codon   The **codons** UAA, UGA, or UAG on **messenger RNA** (mRNA) (see Table 2). Since no **transfer RNA** molecules exist that possess **anticodons** to these sequences, they cannot be translated. When they occur in frame on an mRNA molecule, protein synthesis stops and the ribosome releases the mRNA and the protein.

T

Telomere   End of a **chromosome**. The telomere is a specialized structure involved in replicating and stabilizing linear **DNA** molecules.

Teratogen   Any external agent/factor that increases the probability of congenital malformations. A teratogen may be a drug, whether prescribed or illicit, or an environmental effect, such as high temperature. The classical example is thalidomide, a drug originally prescribed for morning sickness, which resulted in very high rates of congenital malformation in exposed fetuses (especially limb defects).

Termination codon   See **stop codon**

Thymine (T)   One of the bases making up **DNA** and **RNA** (pairs with **adenine**).

Transcription   Synthesis of single-stranded **RNA** from a double-stranded **DNA** template (see Figure 18).

Transfer RNA (tRNA)   An **RNA** molecule that possesses an **anticodon** sequence (complementary to the **codon** in mRNA) and the amino acid which that codon specifies. When the ribosome 'reads' the mRNA codon, the tRNA with the corresponding **anticodon** and amino acid is recruited for protein synthesis. The tRNA 'gives up' its amino acid to the production of the protein.

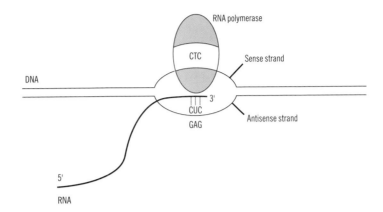

Figure 18. Schematic demonstrating the process of **transcription**. The sense strand has the sequence CTC (coding for leucine). **RNA** is generated by pairing with the antisense strand, which has the sequence GAG (the complement of CTC). The RNA produced is the complement of GAG, CUC (essentially the same as CTC, **uracil** replaces **thymine** in RNA).

Translation

Protein synthesis directed by a specific **messenger RNA** (mRNA), (see Figure 19). The information in mature mRNA is converted at the ribosome Into the linear arrangement of amino acids that constitutes a protein. The mRNA consists of a series of trinucleotide sequences, known as **codons**. The **start codon** is AUG, which specifies that methionine should be inserted. For each codon, except for the **stop codons** that specify the end of translation, a **transfer RNA** (tRNA) molecule exists that possesses an **anticodon** sequence (complementary to the codon in mRNA) and the amino acid which that codon specifies. The process of translation results in the sequential addition of amino acids to the growing polypeptide chain. When translation is complete, the protein is released from the ribosome/mRNA complex and may then undergo post-translational modification, in addition to folding into its final, active, conformational shape.

Translocation

Exchange of chromosomal material between 2 or more non-**homologous chromosomes**. Translocations may be balanced or unbalanced. Unbalanced translocations are those that are observed

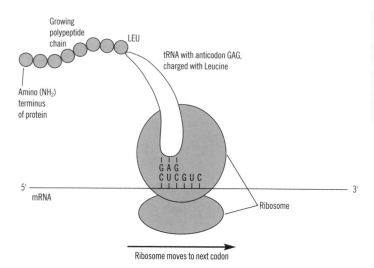

Figure 19. Schematic of the process of **translation**. **Messenger RNA** (mRNA) is translated at the ribosome into a growing polypeptide chain. For each **codon**, there is a **transfer RNA** molecule with the anticodon and the appropriate amino acid. Here, the amino acid leucine is shown being added to the polypeptide. The next codon is GUC, specifying valine. Translation happens in a 5' to 3' direction along the mRNA molecule. When the stop codon is reached, the polypeptide chain is released from the ribosome.

in association with either a loss of genetic material, a gain, or both. As with other causes of **genomic** imbalance, there are usually phenotypic consequences, in particular mental retardation. Balanced translocations are usually associated with a normal **phenotype** but increase the risk of genomic imbalance in offspring, with expected consequences (either severe phenotypes or lethality). Translocations are described by incorporating information about the chromosomes involved (usually but not always two) and the positions on the chromosomes at which the breaks have occurred. Thus t(11;X)(p13;q27.3) refers to an apparently balanced translocation involving chromosome 11 and X, in which the break on 11 is at 11p13 and the break on the X is at Xq27.3

Triplet repeats

Tandem repeats in **DNA** that comprise many copies of a basic trinucleotide sequence. Of particular relevance to disorders associated with **dynamic mutations**, such as Huntington's chorea (HC). HC is associated with a pathological expansion of a CAG repeat within the coding region of the huntingtin **gene**. This repeat codes for a tract of polyglutamines in the resultant protein, and it is believed that the increase in length of the polyglutamine tract in affected individuals is toxic to cells, resulting in specific neuronal damage.

Trisomy

Possessing three copies of a particular **chromosome** instead of two.

Tumor suppressor genes

Genes that act to inhibit/control unrestrained growth as part of normal development. The terminology is misleading, implying that these genes function to inhibit tumor formation. The classical tumor suppressor gene is the Rb gene, which is inactivated in retinoblastoma. Unlike **oncogenes**, where a **mutation** at one **allele** is sufficient for malignant transformation in a cell (since mutations in oncogenes result in increased activity, which is unmitigated by the normal allele), both copies of a tumor suppressor gene must be inactivated in a cell for malignant transformation to proceed. Therefore, at the cellular level, tumor suppressor genes behave recessively. However, at the organismal level they behave as dominants, and an individual who possesses a mutation in only one Rb allele still has an extremely high probability of developing bilateral retinoblastomas.

The explanation for this phenomenon was first put forward by Knudson and has come to be known as the **Knudson hypothesis** (also known as the second hit hypothesis). An individual who has a germ-line mutation in one Rb allele (and the same argument may be applied to any tumor suppressor gene) will have the mutation in every cell in his/her body. It is believed that the rate of spontaneous somatic mutation (defined functionally, in terms of loss of function of that gene by whatever mechanism) is of the order of one in a million per gene per cell division. Given that there are many more than one million retinal cells in each eye, and many cell divisions involved in retinal development, the chance that the second (wild-type) Rb allele will suffer a somatic mutation is extremely high. In a cell that has acquired a 'second hit', there will now be no functional copies of the Rb gene, as the other allele is already mutated (germ-line mutation). Such a cell will have completely lost its ability to control cell growth and will eventually manifest as a retinoblastoma. The same mechanism occurs in many other tumors, the tissue affected being related to the tissue specificity of expression of the relevant tumor suppressor gene.

U

Unequal crossing over

Occurs between similar sequences on **chromosomes** that are not properly aligned. It is common where specific repeats are found and is the basis of many **microdeletion/microduplication** syndromes (see Figure 20).

Uniparental disomy (UPD)

In the vast majority of individuals, each **chromosome** of a pair is derived from a different parent. However, UPD occurs when an offspring receives both copies of a particular chromosome from only one of its parents. UPD of some chromosomes results in recognizable **phenotypes** whereas for other chromosomes there do not appear to be any phenotypic sequelae. One example of UPD is Prader-Willi syndrome (PWS), which can occur if an individual inherits both copies of chromosome 15 from their mother.

Uniparental heterodisomy

Uniparental disomy in which the two **homologues** inherited from the same parent are not identical. If the parent has **chromosomes** A,B the child will also have A,B.

Uniparental isodisomy

Uniparental disomy in which the two **homologues** inherited from the same parent are identical (i.e. duplicates). So, if the parent has **chromosomes** A,B then the child will have either A,A or B,B.

Uracil (U)

A nitrogenous base found in **RNA** but not in **DNA**, uracil is capable of forming a **base pair** with **adenine**.

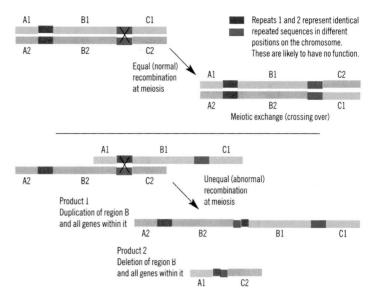

Figure 20. Schematic demonstrating (i) normal **homologous recombination** and (ii) homologous unequal recombination, resulting in a **deletion** and a duplication **chromosome**.

V

Variable expressivity

Variable expression of a **phenotype**: not all-or-none (as is the case with **penetrance**). Individuals with identical **mutations** may manifest variable severity of symptoms, or symptoms that appear in one organ and not in another.

Variable number of tandem repeats (VNTR)

Certain **DNA** sequences possess tandem arrays of repeated sequences. Generally, the longer the array (i.e. the greater the number of copies of a given repeat), the more unstable the sequence, with a consequent wide variability between **alleles** (both within an individual and between individuals). Because of their variability, VNTRs are extremely useful for genetic studies as they allow for different alleles to be distinguished.

W

Western blot   Like a **Southern** or **Northern blot** but for proteins, using a labeled antibody as a probe.

X

X-autosome translocation   **Translocation** between the **X chromosome** and an **autosome**.

X chromosome   See **sex chromosomes**

X-chromosome inactivation   See **Lyonisation**

X-linked   Relating to the **X chromosome**/associated with **genes** on the X chromosome.

X-linked recessive (XLR)   **X-linked** disorder in which the **phenotype** is manifest in **homozygous/hemizygous** individuals (see Figures 21a and 21b). In practice, it is hemizygous males that are affected by X-linked recessive disorders, such as Duchenne's muscular dystrophy (DMD). Females are rarely affected by XLR disorders, although a number of mechanisms have been described that predispose females to being affected, despite being **heterozygous**.

X-linked dominant (XLD)   **X-linked** disorder that manifests in the heterozygote. XLD disorders result in manifestation of the **phenotype** in females and males (see Figure 22). However, because males are **hemizygous**, they are more severely affected as a rule. In some cases, the XLD disorder results in male lethality.

Y

Y chromosome   See **sex chromosomes**

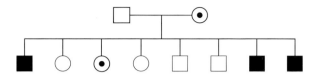

Figure 21a. X-linked recessive inheritance – A. Most X-linked disorders manifest recessively, in that **heterozygous** females (**carriers**) are unaffected and males, who are **hemizygous** (possess only one **X chromosome**) are affected. In this example, a carrier mother has transmitted the disorder to three of her sons. One of her daughters is also a carrier. On average, 50% of the male offspring of a carrier mother will be affected (having inherited the mutated X chromosome), and 50% will be unaffected. Similarly, 50% of daughters will be carriers and 50% will not be carriers. None of the female offspring will be affected but the carriers will carry the same risks to their offspring as their mother. The classical example of this type of inheritance is Duchenne's muscular dystrophy.

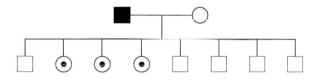

Figure 21b. X-linked recessive inheritance – B. In this example the father is affected. Because all his sons must have inherited their **Y chromosome** from him and their **X chromosome** from their normal mother, none will be affected. Since all his daughters must have inherited his X chromosome, all will be carriers but none affected. For this type of inheritance, it is clearly necessary that males reach reproductive age and are fertile—this is not the case with Duchenne's muscular dystrophy, which is usually fatal by the teenage years in boys. Emery-Dreifuss muscular dystrophy is a good example of this form of inheritance, as males are likely to live long enough to reproduce.

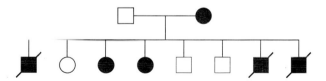

Figure 22. X-linked dominant inheritance. In X-linked dominant inheritance, the **heterozygous** female and hemizygous male are affected, however, the males are usually more severely affected than the females. In many cases, X-linked dominant disorders are lethal in males, resulting either in miscarriage or neonatal/infantile death. On average, 50% of all males of an affected mother will inherit the gene and be severely affected; 50% of males will be completely normal. Fifty percent of female offspring will have the same phenotype as their affected mother and the other 50% will be normal and carry no extra risk for their offspring. An example of this type of inheritance is incontinentia pigmenti, a disorder that is almost always lethal in males (males are usually lost during pregnancy).

Z

Zippering

A process by which complementary **DNA** strands that have annealed over a short length undergo rapid full annealing along their whole length. DNA annealing is believed to occur in 2 main stages. A chance encounter of two strands that are complementary results in a short region of double stranded DNA, which if perfectly matched, stabilizes the two single strands so that further re-annealing of their specific sequences proceeds extremely rapidly. The initial stage is known as nucleation, while the second stage is called zippering.

Zygote

Diploid cell resulting from the union of male and female **haploid gametes**.

5. Abbreviations

ALL	acute lymphoblastic leukemia
AML	acute myelogenous leukemia
Ap-1	activator protein-1
APC	adenomatous polyposis coli
AT	ataxia telangiectasia
ATM	ataxia telangiectasia mutated
BCC	basal cell carcinoma
BCLC	breast cancer linkage consortium
BLM	Bloom syndrome
BMP	bone morphogenetic protein
BRCA1	breast cancer 1 gene
BRCA2	breast cancer 2 gene
BWS	Beckwith–Wiedemann syndrome
B–Z	Bannayan–Zonana syndrome
BZX	Bazex syndrome
CA125	cancer antigen 125
CAL	café au lait
CDH1	cadherin 1
CDK1	cyclin-dependent kinase 1
CDKN1C	cyclin-dependent kinase inhibitor 1C
CHRPE	congenital hypertrophy of the retinal epithelium
CHS	Chédiak–Higashi syndrome
CML	chronic myelogenous leukemia
CNS	central nervous system
CRC	colorectal cancer
CTLA4	cytotoxic T lymphocyte associated antigen 4
DNA	deoxyribonucleic acid
EBV	Epstein–Barr virus
ERCP	endoscopic retrograde cholangiopancreatography
FA	Fanconi anemia
FAMMM	familial atypical multiple mole-melanoma
FAP	familial adenomatous polyposis coli
FBC	full blood count
GI	gastrointestinal
GLUT-1	glucose transporter-1
GTPase	guanosine triphosphatase
HIF-1	hypoxia-inducible factor-1
HLP	hyperkeratosis lenticularis perstans
HNPCC	hereditary non-polyposis colorectal cancer
HPC	hereditary pancreatitis

Genetics for Oncologists

JPS	juvenile polyposis
LFL	Li–Fraumeni-like syndrome
LFS	Li–Fraumeni syndrome
MEN1	multiple endocrine neoplasia type 1
MEN2A	multiple endocrine neoplasia type 2A
MEN2B	multiple endocrine neoplasia type 2B
MIC	morphologic, immunologic and cytogenetic
MRI	magnetic resonance imaging
MSI	microsatellite instability
MYO5A	myosin Va
NBCCS	nevoid basal cell carcinoma syndrome
NER	nuclear excision repair
NF1	neurofibromatosis type 1
NF2	neurofibromatosis type 2
NHL	non-Hodgkin's lymphoma
OCA1	oculocutaneous albinism type I
OCA2	oculocutaneous albinism type II
PI3-K	phosphatidylinositol 3-kinase
PJS	Peutz Jeghers syndrome
PNET	primitive neural ectodermal tumor
PPoma	pancreatic polypeptidoma
PSA	prostate-specific antigen
PSTI	pancreatic secretory trypsin inhibitor
PTCH	patched
PTEN	phosphate and tensin homolog
pVHL	VHL protein
RB	retinoblastoma
RPS19	ribosomal protein S19
RTS	Rothmund Thomson syndrome
SAP	SLAM-associated protein
SCCHN	squamous cell carcinoma of the head and neck
SDHC	succinate dehydrogenase complex, subunit C
SDHD	succinate dehydrogenase complex, subunit D
SLAM	signaling lymphocyte activation molecule
SMOH	smoothened
TGF-β	transforming growth factor-β
TOC	tylosis esophageal cancer
TS	tuberous sclerosis
TYR	tyrosinase
UV	ultraviolet

VEGF	vascular endothelial growth factor
VHL	von Hippel–Lindau disease
VIPoma	vasoactive intestinal polypeptidoma
WAGR	Wilms tumor, Aniridia, Genitourinary abnormalities and mental Retardation
WASP	Wiskott–Aldrich syndrome protein
WT	Wilms tumor
XP	xeroderma pigmentosum
XPV	xeroderma pigmentosum; variant type

6. Index

Genetics for Oncologists

myelodysplasia **83**
MYO5A gene **90**

NBS1 gene **40**
nephroblastoma (Wilms tumor) **102–3**
neurofibromatosis
    type-1 **34–6**
    type-2 **37–8, 65**
nevi, basal cell nevus syndrome (BCCS) **4–6, 105**
NF1 gene **36, 66**
NF2 gene **38, 66**
nibrin/p95 **40**
Nijmegen breakage syndrome **39–40**
non-Hodgkin lymphoma **87, 120**

OCA1/2 genes **105, 112–3**
oculocutaneous albinism (types I/II) **112–3**
oculocutaneous telangiectasia **2–3**
ovarian cancer **60–4, 92–3**

p16, G1 arrest **109**
p53
    ataxia telangiectasia **2–3**
    Li–Fraumeni syndrome **25, 66, 69, 73, 111**
    Peutz–Jeghers syndrome **42**
    retinoblastoma **44**
    BRCA2-related cancers **64**
pancreatic cancer **94**
    hereditary pancreatitis **95**
pancreatic islet cell cancers **28–9**
papillary renal cell carcinoma **98, 100, 101**
    familial **99, 100**
paraganglioma **76–7**
patched *PTCH* gene **5–6, 66, 105**
PCAP gene **97**
Peutz–Jeghers syndrome **22, 41–2**
pheochromocytoma **34**
Philadelphia chromosome **121**
PI3-K proteins **2, 11**
pineoblastoma **65**
poikiloderma atrophicans and cataract **115**
polyposis
    juvenile **22–3**
    see also familial adenomatous polyposis coli (FAP)
primitive neural ectodermal tumors (PNET) **65**
prolactin secretion **28**
prostate cancer **96–7**
    breast cancer-related **61, 63**
prostate-specific antigen (PSA) **63, 96**
PRSS1 gene **94, 95**
PSA *see* prostate-specific antigen
PTCH gene **5–6, 66, 105**
PTEN gene **10–1, 60, 73**
    thyroid cancer **117–8**

RAB27A gene **90**